Shipwrecked

Shipwrecked
New Zealand maritime disasters

GAVIN McLEAN

Edited by Kynan Gentry

Colour plates by Eric Heath

Oratia

FRONT COVER The MV *Rena*, grounded on Astrolabe Reef. Speaking to Radio New Zealand News on the morning of 12 October 2011, author Gavin McLean offered the opinion that 'the *Rena* [was] the worst maritime environmental disaster [New Zealand had experienced].' He was right.
images@visionmedia.co.nz

BACK COVER Painting of the scene of the sinking of the HMS *Orpheus* as it attempted to cross the Manukau Bar, off the Manukau Heads, Auckland.
Auckland Libraries Heritage Collections, JTD-19M-03290-1

HALF-TITLE PAGE Captain Arthur Henry Austin ran the Northern Steam Ship Company's coaster *Gairloch* onto Oakura Reef, north of Cape Egmont, on 5 January 1903 in thick weather. The court of inquiry suspended his certificate for hugging the coast too closely under those conditions. The ship had been built in Scotland in 1884 for the Wairoa trade.
Ref: 1/1-023369-G, Alexander Turnbull Library, Wellington, NZ

TITLE PAGES A tug stands by the liner *Wanganella*, stuck fast on Barrett Reef, near the entrance to Wellington Harbour in 1947. Miraculously, the fine weather held, enabling salvors to refloat the ship after an epic salvage job.
Gavin McLean Collection

The publishers acknowledge the generous assistance provided by Graham Stewart of Grantham House.

Every effort has been made to trace copyright holders and obtain permission for images reproduced in this book. The publisher invites contact from any copyright owners that have not been acknowledged.

Published by Oratia Books, Oratia Media Ltd, 783 West Coast Road, Oratia, Auckland 0604, New Zealand (www.oratia.co.nz).

Copyright © 2019 The Estate of Gavin McLean
Copyright © 2019 Oratia Books (published work)
Copyright in colour plates © 2019 Eric Heath

The copyright holders assert their moral rights in the work.

This book is copyright. Except for the purposes of fair reviewing, no part of this publication may be reproduced or transmitted in any form or by any means, whether electronic, digital or mechanical, including photocopying, recording, any digital or computerised format, or any information storage and retrieval system, including by any means via the Internet, without permission in writing from the publisher. Infringers of copyright render themselves liable to prosecution.

ISBN 978-0-947506-66-7

First published 2007 by Grantham House as *Full Astern!*
This updated and redesigned edition published 2019
Design by Sarah Elworthy

Printed in China

Contents

Foreword *by Kynan Gentry* **6**
Preface to the 2007 edition **7**
Measurements, terminology and abbreviations **8**

Part One
Hazards of the Coasts
1 Taming treacherous coasts **13**
2 Drunk, deranged or dangerous — the human factor **31**
3 Bar harbours — river ports **41**
4 Exposed coasts — roadstead ports **55**

Part Two
The Sail and Steam Era
5 Hot times in cold climes — icebergs and castaways **75**
6 'Terrible engines of destruction' — collisions **87**
 Colour section 1 **following page 88**
7 'A thrill of horror throughout the Empire' — fire **99**
8 Two of the worst — the *Wairarapa* and the *Penguin* **109**

Part Three
Foul Play
9 Fraud, vandalism and terrorism **127**
10 Casualties of war **137**

Part Four
The Modern Era
11 Conference Lines casualties **149**
12 Floating Jonahs and hoodoo ships — unlucky ships **161**
13 Coasters in crisis **175**
 Colour section 2 **following page 184**
14 Close calls — successful salvages **189**
15 'Rocks ahead! ... Rocks astern!' — passengers in peril **199**
16 From horror to heritage **213**
17 Shipwrecks in the twenty-first century **225**

Further resources **235**
Select maritime chronology **236**
Notes **240**
Index **245**

Foreword

Getting a book published is a challenging undertaking. Aside from the mere process of researching and writing, as an author you have to conceive of a topic that is both original in what it may add to our collective knowledge, while also commercially viable for potential publishers. It's thus understandably rare that an author gets an opportunity to produce a second edition, and rarer still a third. Indeed, a third edition suggests that the book in question is a 'go-to text' and is on the way, dare I suggest, to becoming a classic.

So it was understandable how quietly proud Gavin McLean was when in early February Oratia Books approached him with the idea of publishing that elusive third edition. A number of his books had seen multiple printings or second editions, but to my knowledge, this was to be his first third. Despite failing health, he immediately got to work making notes on updates, omissions and new material. After all, the period since the 2007 edition had witnessed the saga of the MV *Rena*, the largest ship ever wrecked in New Zealand waters, and the country's worst environmental maritime disaster.

Unfortunately, Gavin did not live to see the book come off the presses. In the more than 20 years I had known him, that had been something of a ritual; the cracking open of the publisher's courier package of 'author copies' typically sent out as soon as the books land in the warehouse. Flicking through the pages, it was the smell of new books that captured his imagination most but, satisfied with the final product, it would usually then be added to the 'trophy shelf' of his published works, as he settled back and continued work on yet another tome. Initially this was an unassuming shelf built into his computer desk, but by the early 2000s, creaking under the strain, the trophy shelf was upgraded to a full bookcase. It needed to be. In all, my count lists Gavin as sole or co-author of a staggering 53 books, and editor of a further 10. Few New Zealand authors have made a greater contribution.

While he was also expert in government and organisational history, heritage, and a fierce advocate of public history, Gavin's true passion was maritime history, with almost half of his published output focusing on some aspect of New Zealand's seafaring past. His focus was wide ranging as he pursued topics as broad as vessel and shipping company histories, chronicles of port towns, collections of nautical yarns and even a history of the Lyall Bay Surf Club—not to mention shipwrecks.

As Gavin noted in the preface to the 2007 edition, the book began life in 1991 as *New Zealand Shipwrecks and Maritime Disasters*. More than simply a rich social history of shipwrecks in New Zealand, it includes everything from the political and economic development of maritime industry—so important to the growth of a small island nation—to the role of private enterprise and its testy relationship with government and regulation. All through these pages the human face of shipping (and ultimately the consequences of wrecks) remains central. That reflects what was so unique about Gavin's talent: his genius for storytelling.

In completing this book, such variables have been ever present in my mind. Where possible I've sought to leave light footprints, careful not to impinge on Gavin's own voice, while at the same time seeking to interpret his notes and ideas for the book's development. Many people have aided in this, though particular thanks need to go to Deb Hill for proofing, Mike Pryce for being able to answer seemingly any technical question I could throw at him, and to the many copyright holders who offered the use of their images.

Finally, thanks again to Peter Dowling from Oratia Books, and editorial director Carolyn Lagahetau and her team for production.
Kynan Gentry

Preface to the 2007 edition

Shipwrecks excite. Tales of calamity at sea run through the ancient threads of both our principal cultures, whether they are Europe's *The Odyssey, Ulysses* and *Jason and the Argonauts,* or the legends of Polynesia's *Araiteuru* waka. Fireside yarn-spinners knew that they could rely on a good wreck story to deliver everything a novelist could desire: cruel fate, human drama, bravery (or, even better, abject cowardice), courtroom confrontations and legal shenanigans. Some even threw in a touch of cannibalism. There was no such drama on the only wreck I have witnessed — the grounding of the trawler *Sea witch* on the Oamaru foreshore after a rope tangled the propeller — but the incident brought home the drama inherent in even a bloodless stranding, as crowds gather and people and equipment battle tide and wind.

There are many fine books on New Zealand shipwrecks. Most recount single incidents — the *Wahine, Mikhail Lermontov, Tararua, General Grant* etc — and since 1936 there has been that indispensable reference tool, Dunedin firefighter C.W.N. Ingram's labour of love, *New Zealand Shipwrecks*. At the time of writing Ingram's book is being updated and revised for publisher Hodder Moa, after being out of print for over a decade.

But such books deal only with the most notorious wrecks, and usually only with ships that became total losses. In this book I offer a sampling of some of the more important or calamitous wrecks, total or otherwise, linking sometimes chronologically distant accidents in a thematic order that I hope will illuminate the story of what we should see as less a tally of misfortunes than as blips in the process of improving safety at sea. As wrecks are only part of the maritime story, the first section of the book summarises regulatory and technological developments over time and our responses to the basic port types, rivers and roadsteads.

This book began life in 1991 as *New Zealand Shipwrecks and Maritime Disasters,* part of Grantham House's 'New Zealand Tragedies' series. It was reprinted in 1995, but has been out of print for many years. It was, therefore, a pleasure to revise and expand the book in response to enquiries for a new edition. I have completely rewritten the text, expanded most chapters, added new wrecks and incidents, updated earlier material in the light of further research, included many new illustrations and provided a new final chapter on shipwreck history and heritage.

In 1990/1 Colin Watson, Ken Scadden and Emmanuel Makarios from the Wellington Maritime Museum and Gallery (now the Museum of Wellington City & Sea), successors to Jack Churchouse, who first set aside material for this book, gave me free rein of their photographic collection, and Ken and Bob McDougall from the New Zealand Ship & Marine Society read through the manuscript. For this book I would like to thank Joan McCracken and Heather Mathie from the Alexander Turnbull Library for their help in providing photographs, Captain Mike Pryce and Ian Farquhar for the generous loan of photographs, and Kynan Gentry for digitally remastering some sorry originals.

Finally, thanks again to publisher Graham Stewart from Grantham House and Lorraine Olphert for her editing.
Gavin McLean

Several *Ronas* have been wrecked in New Zealand waters. This one was lucky. The Colonial Sugar Refining Company's big freighter *Rona* hit Flat Rock in the Hauraki Gulf on 26 June 1922 but was refloated and repaired.
New Zealand Herald

Measurements and terminology

I have metricised ship dimensions (1 foot= 0.3038 metre) and land distances (1 mile= 1.6 km), but distances at sea are still expressed in nautical miles (1 nautical mile = 1.852 km). I have converted cargo weights to tonnes, but not ship 'weights', since the most common measurement of merchant ship size, gross registered tonnage (grt) is a measurement of volume, not weight (100 feet = 1-ton gross), and is still expressed in tonnes.

Inflation has rendered any simple comparison between pre-decimal pounds and dollars meaningless. One 1967 pound equalled two 1967 dollars. The Reserve Bank of New Zealand's CPI inflation calculator http://www.rbnz.govt.nz/ provides a useful price comparison.

Barque	Sailing vessel of three or more masts, square-rigged on all but the mizzen (rear) mast.
Barquentine	Sailing vessel of three or more masts, fully square-rigged on the foremast and fore-and-aft rigged on the other masts.
Brig	Two-masted sailing vessel, square-rigged on both.
Brigantine	Two-masted sailing vessel, square-rigged on only the foremast.
Bowsprit	The spar projecting from a ship's bow (front).
Coamings	The built-up surrounds to a ship's hatches.
Cutter	A small, single-masted sailing craft; also used to describe a large open boat carried aboard a ship.
Fore-and-aft rig	Sails set along the central bow-to-stern line of a ship; typically carried on schooners, ketches and cutters.
Forecastle	The built-up structure at the bow of a ship.
Intercolonial	Obsolete term for trans-Tasman.
Jury rig	Improvised rig.
Keelson	Strengthening timber attached to a ship's keel.
Ketch	Two-masted fore-and-aft rigged sailing vessel, similar to a schooner, but with a mizzen mast shorter than the foremast.
Lee shore	A shore toward which the wind blows and toward which a ship is likely to be driven.
Lose way	Slow astern.
Mainmast	The tallest mast in a ship, next behind the foremast.
Miss stays	To fail in going about from one tack to the other, as a result of which the ship gets its head to the wind, comes to a stand and begins to fall off on the same tack.
Mizzen	The last mast in a sailing vessel.
Port	Left.
Poop	The highest and aftermost deck of a ship.
Prize crew	Crew put aboard a captured ship.
Quarterdeck	The deck raised above the main deck and running towards the stern.
Schooner	Sailing vessel of two or more masts, fore-and-aft rigged (topsail schooners carried some square sails on the foremast).
Scow	Small, flat-bottomed sailing vessel (later some were motorized).
Ship (full-rigged)	Sailing vessel of three or more masts, carrying square sails on all.
Shrouds	Rigging supporting the mast, part of the standing rigging.
Standing rigging	The fixed ropes/wires that support a vessel's main spars, masts, bowsprit and jib-boom.
Starboard	Right.
Stay	A brace consisting of a heavy rope or wire cable used as a support for a mast or spar.
Tack	To turn a sailing vessel by moving its bow forwards, then past the source of the wind.
Touch	To strike the seabed or a submerged object.

Abbreviations

AB	Able-bodied seaman
AJHR	*Appendix to the Journals of the House of Representatives*
NZSCo	New Zealand Shipping Company
OS	Ordinary seaman
TAIC	Transport Accident Investigation Commission

As the steamer *Wakatu* sailed from Wellington to Kaikoura in thick fog early on the morning of 6 September 1924, an unusually strong current threw it high onto the beach on the Blenheim side of Waipapa Point, Marlborough. The ship was under the command of its regular master, Captain D. Robertson, who was exonerated by the court of inquiry, which blamed the weather conditions. The ships' resting place so far up on the beach greatly simplified the recovery of the cargo but made salvage of the ship, which was holed and buckled near the stern, impossible. The *Wakatu* was left to rust away. The iron, screw steamer of 95 tons had been built at Nelson in 1879 and was owned by the Wellington-based Wakatu Shipping Company.

ABOVE Gavin McLean Collection **LEFT** Graham Stewart

Hazards of the Coast

PART ONE

1
Taming treacherous coasts

It had been a dark and stormy night. On the morning of a wild April day in 1968, two businessmen were chatting in a cabin aboard one of the world's most modern passenger vessels while waiting for it to berth in the capital. The 9000-ton vessel was powerful, manoeuvrable, and carried every modern aid to navigation. Forty-nine-year-old Frank Penman was talking to his boss when he saw 'an ugly sight of foaming violent sea with the wind sweeping the spray in spirals towards the forward end of the vessel'. He looked away, then the ship lurched, 'propelling my managing director from his bed across the cabin to uncontrollable impact with myself and the furniture …'[1] As the men picked themselves up, they saw green water covering the window. That was the beginning of a day neither man — nor the country — would forget.

Their ship was the *Wahine,* the inter-island ferry whose destruction imprinted itself on the national memory. 'Wahine Day', as it came to be called, reminded New Zealanders that the sea, their commercial highway for centuries, sometimes imposes severe tolls on users. Seafarers have known this for millennia. Homer's Greeks, Hawaii's Polynesians or today's container ship jockeys, all know to treat the sea with respect.

Neptune, Tangaroa, call it what you like, can kill. In fact, it has claimed well over 2000 ships and fishing vessels off our coasts alone. We will never know the names of the first victims. What we do know is that Eastern Polynesian colonisers settled here permanently (there may have been earlier reconnaissance voyages) over 800 years ago. Their descendants, Maori, have handed down legends that include references to early maritime casualties. One was the waka atua ('canoe of the gods') *Araiteuru,* which left an unknown land called Taitewhenua long ago. After stopping at Hawaii, the traditional setting-off point for the later migrations to New Zealand, the ancient voyagers sailed south, making landfall at Turanga near Gisborne. There they planted the

OPPOSITE In 1769 de Surville's ship *Saint Jean Baptiste* came close to grief off the Northland coast.
Gavin McLean Collection

PAGES 10–11 In September 1913 a storm in the Tasman Sea damaged the Norwegian barque *Okta* (1110 tons, 1874), forcing it to seek shelter at Bluff. Unfortunately, because he had not intended to call at New Zealand, Captain C. Duus had no charts for the port. He threaded his way through the hazards of Foveaux Strait only to hit rocks at the entrance to Bluff.
Ref: PAColl-4286-1, Alexander Turnbull Library, Wellington, NZ

first kumara, before heading further south to Te Waipounamu (the South Island).

Off North Otago luck ran out for the *Araiteuru*. Near Shag Point the current swept the waterlogged waka on to Araiteuru (Danger Reef), scattering its cargo of kumara, taro and gourds along the beach to create the Moeraki Boulders. Maori memorialised the wreck by naming local features after it. *Araiteuru*'s broken mast and sail, turned to stone, are said to be visible near Matakaea (Shag Point) and the great waves that wrecked it are represented by the range of hills running parallel to the coast. Many carry crew members' names.

Another southern tradition explains how greenstone (pounamu) got to Te Waipounamu's West Coast. The Pounamu people lived in Hawaii but, fearing the tribes named Mata (flint) and Hoanga (grindstone), they fled south in the waka *Tairea*. Off the West Coast the *Tairea* was wrecked and the crew turned into pounamu, taking refuge from their enemies in the rivers and waterfalls where greenstone is now found.[2]

The documentary trail became clearer after 1642 when the Dutch navigator, Abel Tasman, arrived with the ships *Heemskerck* and *Zeehaen*. He fled after only a brief encounter, leaving the task of putting the archipelago 'on the map' to James Cook, who during three epic voyages between 1769 and 1777 charted the coast accurately. Exploring uncharted coasts was dangerous. Cook clawed his way up and down the coast cautiously, taking soundings frequently and posting his sharpest-eyed men as lookouts. That disaster was only a puff of wind away became clear in December 1769, when French explorer Jean-Francois Marie de Surville almost went ashore in Doubtless Bay, Northland. A storm had carried away the *Saint Jean Baptiste*'s rudder and de Surville was lucky to lose only anchors. Two were recovered late last century and are on display at the Museum of New Zealand and the local museum as memorials to New Zealand's first documented maritime accident.

Europeans kept clear of New Zealand until the 1790s when New South Wales entrepreneurs discovered the islands' huge stocks of fur seals. Between 1790 and the early 1840s they ruthlessly exploited the seals and then whales, with devastating effects on the hapless mammals … and sometimes on themselves. Sailing virtually uncharted waters for months at a time was never safe, and it is not surprising that the first ship known to have been wrecked here was a store ship for sealers.

That ship, the *Endeavour* (not Cook's ship), limped into Dusky Sound in October 1795, its pumps struggling to keep its decayed timbers afloat. Sealing parties had been working from the sound since 1792. By the time the *Endeavour* dropped anchor, a house and a small jetty had been erected and an earlier party had even half-built New Zealand's first European-designed ship, a 60-ton schooner. The *Endeavour* was surveyed and declared unseaworthy and days later hit a rock, further damaging its hull. When a cable chafed through early in November, the old ship was run ashore, beached and stripped. The arrival of another ship, the *Fancy*, brought the sound's European population to over 200. Its crew helped to finish the schooner (which was named *Providence*) and to convert the *Endeavour*'s longboat into a deep-sea craft (the *Assistance*). All three made it back to New South Wales early in 1796, carrying most of the sealers and seafarers. An American whaler picked up the remainder in 1797.

Maori attacked and destroyed some ships. Both sides had exchanged occasional blows since Tasman's day and the situation was certainly not helped by sealers and whalers kidnapping Maori for service as pilots, deliberately or unintentionally flouted local customs, and took goods and resources without paying. In 1808 the crew of the schooner *Parramatta* paid dearly for wounding three Bay of Islands Maori while skipping town without paying for a cargo. The wind changed and blew the *Parramatta* back onto the rocks at Cape Brett, where all were killed.

The first deliberate wrecking had occurred nearby two years earlier, that of the brig *Venus*. Mutineers, including the obese Charlotte Badger, seized the ship at Port Dalrymple, Australia, in June 1806 and took it to the Bay of Islands where local Maori captured the brig, slaughtered all but one of the crew and stripped the hull of everything useable. In March 1816 Tokomaru Bay Maori boarded the American brig *Agnes*, killed most of the crew and plundered and later burned the ship.

The most notorious episode was the loss of the *Boyd*, described by Wade Doak in *The Burning of the Boyd* as a 'saga of culture clash'.[3] The English brig came to Whangaroa to load the kauri spars and was the third ship sent there by the same entrepreneur. Previous ships had been well treated by the local tribes. The last visit had been followed by a devastating epidemic which had killed many Maori, causing them to believe that its captain had placed a curse on them. Future visitors would not be so welcome.

The *Boyd* dropped anchor in December 1809. During the voyage, Captain John Thompson, an insensitive brute, had ill-treated the Whangaroa chief 'George' (Te Ara) who was working as a seaman, flogging him for ignoring orders and subjecting him to other indignities. Te Ara had been visiting Sydney and was returning with other Ngati Pou, all of whom were spoiling for vengeance by the time the *Boyd* reached New Zealand. Three days after its arrival they lured most of the crew away on a timber-cutting expedition and killed them. They returned to the lightly manned ship after nightfall and

Walter Wright painted 'The Burning of the *Boyd*, Whangaroa Harbour, 1809' almost a century after the event. It was an inaccurate representation (the attack was launched at night, not in day light) and has been termed 'documentary racism', but Wright painted what Europeans wanted to see and his work still influences our views.
New Zealand Herald

killed all except four who had shown kindness: cabin boy Thomas Davis, Anne Morley, her baby Betsey Broughton and the second mate. The child and the second mate later died. In all, about 70 were killed, most also being eaten. Although a neighbouring chief, Te Pahi, intervened, he failed to save many.

The following day the Maori dragged the *Boyd* closer to the village and started stripping it of useful items such as nails and fittings. While they were doing this, someone unwisely struck a flint on an opened keg of gunpowder, killing several people, wrecking the *Boyd* and starting a fire that burned the ship to the waterline. 'We found the wreck in shoal water at the top of the harbour. Not far from the entrance', wrote Alexander Berry, the head on an expeditionary force a few weeks later.

> A most melancholy picture of wanton mischief. The natives had cut her cables and towed her up the harbour till she grounded, and then set her on fire, burning her to the water's edge. In her hold were the remains of her cargo; coals, salted seal skins and planks. Her guns, iron standards, etc were lying on top, having fallen in when her decks were consumed.[4]

Nearby on the beach were 'the mangled fragments and fresh bones of our countrymen, with tooth marks still upon them'.[5] This was not the end of the killing. Berry wrongly accused chief Te Pahi of the massacre (or 'massacree' as one sensational broadsheet called it) and launched a punitive strike against his island pa, killing many innocent Maori, including Te Pahi himself. For years the *Boyd* cast a baleful shade over European trade and settlement

Ports varied in the facilities they could offer. Along the coast of the Far North and its east coast, steamers used open roadsteads until well into the twentieth century. Here, sacks of gum are being lightered out to a steamer at Ahipara. Masters always kept a 'weather eye' out on these occasions.
Photographs of Northland, Ref: 1/1-006272-G, Alexander Turnbull Library, Wellington, NZ

> 'The New Zealand death' became something of a colloquialism for drowning, with early drowning rates being very high by modern standards. Fewer people knew how to swim, and as they used small boats much as we now use cars, accidents were common.

intentions. 'Touch not that cursed shore lest you/ The Cannibals pursue', a pamphlet warned.

Other ships lost as the direct result of Maori hostility were the whaling brig *Dragon* in 1833 and the French whaler *Jean Bart* at the Chatham Islands in 1838. The latter became a total loss, captured and burned at Ocean Bay after a violent fight initiated by the Maori response to the Frenchmen's punishment of Maori thieves. All 40 aboard the ship died. The Europeans later exacted revenge by burning the Maori village.

The 'Harriet Affair' of 1834 was another sensation. When the barque *Harriet* was blown ashore near Rahotu on the Taranaki coast, the survivors were first plundered by Taranaki Maori and then later attacked by Ngati Ruanui, who killed 12 of them. In a famous incident, whaler wife Betty Guard's tortoiseshell comb saved her life by taking the main force of a tomahawk blow that left several comb teeth in her skull for the rest of her life. After Ngati Ruanui released Jacky Guard and others to obtain a ransom, they sailed to Sydney, where the incident had become a sensation. Guard returned with HMS *Alligator* and the schooner *Isabella*, which bombarded the Maori positions, landed troops and rescued the hostages.

As historian James Belich observes, the *Boyd* and *Harriet* incidents attracted attention because they were the exceptions, not the rule. Interaction between Maori and European was generally positive and profitable, since each wanted trade. By the time the first New Zealand Company immigrant ships reached Wellington in 1840, ships had little to fear from Maori. Natural causes and human failings were the threat now. Indeed, 'the New Zealand death' became something of a colloquialism for drowning, with early drowning rates being very high by modern standards. Fewer people knew how to swim, and as they used small boats much as we now use cars, accidents were common. People also fell off wharves or slipped into rivers with surprising frequency. Early Dunedin newspapers carried more than one report of watermen finding rats munching on corpses under the first rickety jetty.[6] Boat capsizes killed many more — in 1850 one Port Chalmers rowboat claimed four lives in two accidents. Even as late as 1866 accidents such as this one were common: 'During some squalls at Port Chalmers last Wednesday two boats capsized,' the *North Otago Times* noted. 'In one were Captain Ridley and one of his sons, who both perished; and in the other two fishermen and a settler named William Geary, the latter of whom was drowned and the other two rescued by Pilot Patten and a volunteer crew. Quite a gloom has been cast over the inhabitants of the port from the melancholy occurrence.'[7] As late as 1928, drowning claimed more New Zealand lives than road accidents.

Since little of the coastline had been charted in detail, early mariners had to feel their way, and several ships ended their days on 'uncharted' rocks. In 1867, for example, the Marine Department drew attention to the unhappy fate of the 811-ton ship *Cambodia,* bound in ballast for Howland's Island, but calling at New Zealand for stores, went aground in the Manukau Harbour. The department was critical of the master who 'deserted her somewhat prematurely, since she was afterwards taken into port and has since been repaired' but blamed the accident on a faulty general chart that had 'Auckland' printed on the west side of the island, misleading the captain into thinking he was approaching a port with easy access. 'It is to be regretted that the name of the publisher of the chart was not given at the inquiry.'[8]

Slowly but surely, the coastline was made safer for seafarers. Early harbour works were primitive — a signal station here, some buoys there, and usually precious little else to guide the mariner. Much was left to private enterprise, although the provincial governments established harbour departments in the 1850s. Richard Driver of Otago felt particularly hard done by, complaining about shoddy beacons and buoys ('mostly worm-eaten and sunken or had gone adrift') and the fact that the government's refusal to pay for oil meant that he seldom lit the lamp at the flagstaff.[9] A month later, the immigrant ship *Maori* almost hit rocks, partly because Driver had been unable to secure a crew *for* the pilot boat.

Each port had its own peculiarities, but the colony's harbours fell into three broad categories: the major ports, such as Auckland, Wellington and Lyttelton, which were safe natural anchorages and which required little human improvement beyond the provision of wharves and warehouses; the river ports such as Kaiapoi or Greymouth, which provided safety after work on training walls and dredging; and open roadsteads such as Oamaru or New Plymouth, which required extensive harbour works before they could be

Lighthouses were literally beacons of hope, guiding shipping to the major ports and warning them of the presence of dangerous reefs, rocks and headlands. In 1864 Dunedin builder Hugh Calder built the Taiaroa Head lighthouse at the entrance to Otago Harbour from stone quarried on the site. These days tourists come from around the world to see the albatross colony, but 100 years ago Taiaroa Head was home to the lighthouse keepers, the harbour board signal station staff, pilots and boat crews and even a coastal artillery contingent.
Ref: PA1-o-287-47, Alexander Turnbull Library, Wellington, NZ

made safe for shipping. The hazards posed by early river ports (Hokitika and Greymouth, for example) and open beach ports (for instance, Oamaru) will be illustrated in Chapters 3 and 4.

By the 1870s and 1880s navigation had become safer. Competition from railways (especially in the South Island) weeded out the smallest and most hazardous ports. The harbour boards — local government organisations — which spread like a bureaucratic rash during the 1870s and early 1880s, splurged on lavish port development work, sometimes over-reaching themselves financially, but making their anchorages safer.

As a result, by the end of the Victorian era the number of shipwrecks had declined. Ports were better, ships were powered, and more services were available to mariners in trouble. The statistics were impressive: between 1860 and 1900 2166 people died at sea, an average of 54 a year; between 1900 and 1940 fatalities dropped to 608, an annual average of 15.[10]

Early steamers were not much better than sailing vessels, being underpowered and unreliable, but from the mid-1870s the more efficient compound engine gave the steam ship the edge over the sailing ship. Small steamers had been acting as tugs since the 1850s, but during the 1870s and 1880s harbour boards started to provide adequate towage and salvage services.

During the same period, central government did its inadequate bit to make the coastline between ports safer. Following the abolition of the provinces in 1876, the Marine Department (which had replaced the earlier Marine Board in

By funding lighthouses, lighthouse tenders and depots for castaways, the government made New Zealand's seas much safer during the last decades of the nineteenth century. The best-known of the 'government steamers', as these multi-purpose ships were known, were the 1876 duo, the *Hinemoa* (pictured) and *Stella*, forever linked to long-serving masters such as Captain John Fairchild and Captain John Bollons. The *Hinemoa* was also designated the government and parliamentary yacht, designed for transporting parliamentarians to Wellington. They also carried the governors on summer junkets to the Fiords and the sub-Antarctic islands, but most of their work was less exalted, laying cables, inspecting moorings, surveying the coast and landing supplies for the lighthouses.
Gavin McLean Collection

1866) took control of the ports and anchorages not administered by harbour boards and sped up lighthouse construction. The first permanent lighthouse was completed at Pencarrow, near the entrance to Wellington Harbour, in January 1859. Often such work was reactive, as lighthouses tended to spring up like maritime mushrooms in the wake of disasters. The tower erected at Waipapa Point, Southland, in 1884, three years after the Union Company liner *Tararua* was wrecked with the loss of 131 lives, was typical.

The Marine Department's other improvements included the lighthouse tenders *Stella* and *Hinemoa,* which reached New Zealand in 1876. For decades these graceful little steamers transported the building materials for lighthouses to isolated sites, serviced them and supplied their keepers, searched for missing ships, and restocked the depots built on the sub-Antarctic islands for castaways. Many seafarers owed their lives to the government steamers.

What happened when a shipwreck occurred in colonial days? Then, as now, the first priority was to save lives. If the vessel was wrecked on an isolated part of the coastline, they had to fend for themselves. If the wreck happened near a major port, rescue was normally faster. Several towns formed rocket brigades, hardy volunteers who used rocket guns to fire lines to a ship. These crude but effective guns saved hundreds of lives.

Once the people were safe it was time to worry about the ship and cargo. Many ships were refloated and repaired; Chapter 14 describes some memorable salvage jobs. In many cases, however, salvage was impossible and the ship was 'abandoned to the underwriters'. The insurers now took over, putting the wreck and cargo up for auction. Bidders had to gamble. Tens of thousands of pounds worth of ship and cargo could be knocked down for few hundred pounds. If their luck held, and the seas remained calm, the salvors made a huge profit; very occasionally, they even refloated the ship. But they could also lose their money. One storm was all it took to break up the strongest hull, scattering cargo and wreckage for kilometres.

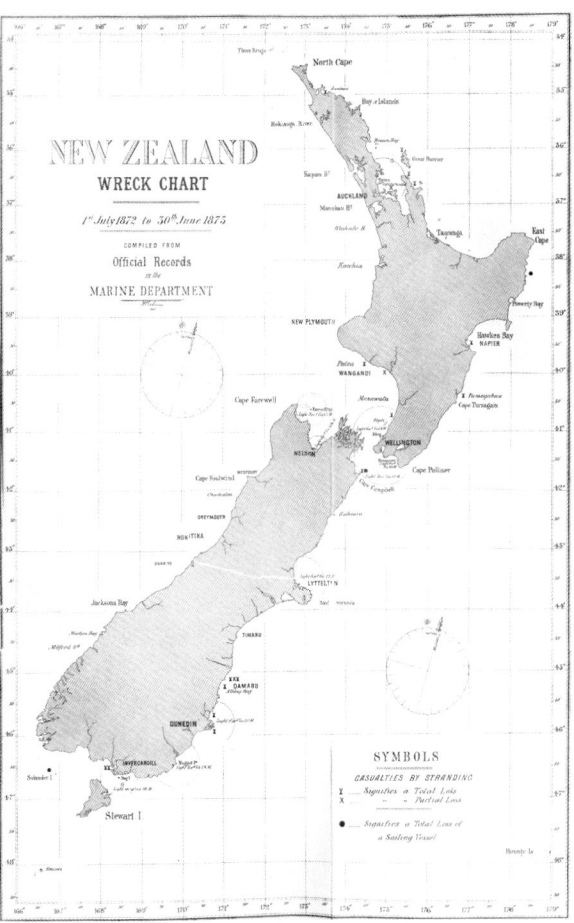

Every year the Marine Department printed a wreck chart showing the location of the previous year's casualties.
Appendix to the Journals of the House of Representatives

Meanwhile, officialdom examined the entrails, literally in the case of coroners. In 1863 Parliament passed the Enquiry into Wrecks Act. This empowered the principal customs officer (or any other person deemed suitable by the governor) at the port nearest to a shipping casualty to hold a preliminary inquiry. Then, as now, preliminary inquiries sought to establish the cause of the accident, not to blame or punish. If the person conducting the preliminary inquiry thought a more formal inquiry was warranted, he (it was always a he then) applied for one. Formal

Seafarers call death 'crossing the bar', a recognition of the dangers of entering such ports. These photographs were taken aboard the barque *Manurewa* (upper photograph) as it was being towed across the Whangape bar by the small tug *Ohinemuri*, whose mast and funnel are just visible in the lower image.
TOP Gift of Wellington Museums Trust, New Zealand Maritime Museum (2012.0.3252) **ABOVE** Gift of Wellington Museums Trust, New Zealand Maritime Museum (2012.0.3253)

investigations were normally presided over by two justices of the peace or a resident magistrate. Six years later the Act was amended to enable justices or magistrates to cancel or suspend seafarers' certificates. In 1877 the provisions of the act were incorporated in the Shipping and Seamen's Act, and the Marine Department was made the administering body. Later amendments to the act altered details but left the main features intact.

Formal inquiries were held in about a quarter of nineteenth-century casualty cases. The presiding magistrates or justices of the peace were usually assisted by one or two professional assessors — often mariners. Hearings could last for days, reported fully by the papers. The courts could exonerate mariners by returning their certificates or could impose penalties, which ranged from reprimands with no suspension of certificates, through to suspension of certificates and imprisonment. In the twentieth century the Marine Department, under the Shipping & Seaman Act 1903 (and later amendments) held a preliminary inquiry. If it felt that the case warranted it, it recommended that the minister hold a formal inquiry.

Court findings sometimes led to major change. As we have seen, the inquiry into the wreck of the steamer *Tararua* in April 1881, with the loss of 131 lives, resulted in the erection of a lighthouse at Waipapa Point, Southland.

The loss of the *Penguin* in February 1909 (see Chapter 11), at the cost of 72 lives, eventually produced the Karori Rock automatic light. The sinking of the coaster *Ripple* off Cape Palliser in 1924 with all hands, brought new regulations for the mandatory installation of radio equipment aboard home trade ships.

Built at Linthouse, Glasgow, in 1902, the *Opua* ran aground in fog at Tora, south-east of Palliser Bay, on 2 October 1926, while on a voyage from Gisborne to Wellington. The fog was thick at the time, but the court of inquiry found that Captain E.H. Fowler contributed to the accident by setting too fine a course from Castlepoint to Cape Palliser and in having no lookout with a knowledge of the coast during the second officer's watch. It also criticised the second officer for not summoning Fowler when visibility deteriorated. Both officers were ordered to pay for the cost of the inquiry.
TOP Brian M. Harmer **ABOVE** Gift of Wellington Museums Trust, New Zealand Maritime Museum (2012.0.764)

Towards safer coasts

From the end of the nineteenth century shipwrecks declined. By the late 1890s casualties were mostly minor, more often the result of machinery breakdown or berthing accidents — embarrassing to the master's prestige and expensive for the shipping company, but seldom life threatening. The dramatic wrecks rendered so heroically in the old Victorian engravings were as rare as moa's teeth. Most mariners expected nothing worse than a simple bump to the bowsprit or a few hours' drifting about when a shaft snapped.

Radio and radar greatly improved safety. Radio had been installed in several passenger ships after the *Titanic* sank in 1912, three years after New Zealand regulations made its installation mandatory in passenger vessels over 45 metres long. Unfortunately, many shipowners resisted upgrading their smaller ships. Parallel improvements in the meteorological service enabled masters to receive more accurate weather forecasts. Radio beacons, which enabled ships to pinpoint their location accurately, were commissioned at Cape Maria Van Diemen and Baring Head in 1926 and 1937 respectively. By the outbreak of World War II similar beacons had been established at seven other stations. By this time most large trans-Tasman ships and coasters carried radio direction-finding equipment.

The use of radar, the electronic device that enabled mariners to 'see' in the dark or in fog, became widespread during World War II. Sets appeared on New Zealand ships during the 1950s but too slowly to satisfy the officers' union, the New Zealand Merchant Service Guild. Some

Eventually, artificial breakwater ports such as Oamaru's, seen in this Burton Brothers image circa 1880s, replaced open roadsteads.
Ref: 1/2-C-22767-F, Alexander Turnbull Library, Wellington, NZ

owners heeded the call but others did not, pleading poverty. Although the inquiry observed that radar might have prevented the wreck of the *Holmbank* off Banks Peninsula in 1963 (see Chapter 13) it was 1970 before its fitting was universal, even aboard 1000-ton ships. It is not a complete guarantee of safety, as accidents showed, but radar makes a huge difference, especially for night navigation within port limits.

The gutting of coastal shipping in the 1960/70s weeded out the smaller ships and outports. Westport, traditionally a dangerous harbour, is now managed by a subsidiary of its main client, a cement company, and offers advanced satellite weather prediction and automatic depth sounding advice for port users. In any case, today's modern coastal freighters are elaborately fitted out with radar, radio and satellite navigation devices and their crews are better trained, with bridge resource management (BRM) courses being common for members of the multi-skilled integrated ship management regimes brought in when crew numbers were cut late last century. Twenty-first century accidents have included some strandings in the Manukau Harbour and collisions by the *Spirit of Resolution* (2005) and *Westport* (2003) with bridges near the Onehunga Wharf, but such incidents have bent steel, not bones. Nowadays, casualties are more likely to be fishing vessels or recreational craft — the occupants of the latter depressingly often ill-equipped or 'three sheets to the wind'.

If a ship does get into difficulties — and machinery failures in confined shipping channels are not uncommon — the crew can call on more sophisticated assistance. In the 1970s and the 1980s even the secondary ports invested in powerful tractor tugs. Backing them up are the resources of the defence forces (whose naval and air force helicopters are an integral part of modern search and rescue operations), rescue and air ambulance helicopters, police launches and a growing fleet of small coastguard vessels.

TUGS

Tugs are familiar to most city dwellers. These powerful little ships play an important part in port life, berthing and unberthing ships, towing barges and on occasion pulling ships off sandbanks or reefs.

Early colonial ports lacked tugs until the 1870s and the 1880s when the larger harbour boards acquired one or two. The side-wheel paddle *Koputai* (top image) was a typical tug of its day (1876). It had been built for a Dunedin company but in the 1880s, dissatisfied with the service offered, the Otago Harbour Board (OHB) bought the ship and operated it until World War I.

When it came time to retire the *Dunedin*, the *Koputai*'s replacement, the OHB ordered the *Otago*. The *Otago* of 1956 (lower image) handled container ships and big bulk carriers. In 1974/75, therefore, OHB staff built two new tugs, the *Rangi* and *Karetai*. They were smaller but more economical (five-man crews), powerful (30 tons bollard pull) and manoeuvrable, and their compact superstructures let them get under the knuckles and flared bows of big modern ships.

The 28–35 tons bollard pull was considered adequate for ship handling until the 1990s, when bigger container ships and ungainly new-generation car carriers forced a rethink. These leviathans are so big and present so much 'windage' that very powerful tugs are needed to manoeuvre them or to hold them against a wharf in high winds. The latest *Otago* was built for Port Otago Ltd at Whangarei in 2003. The highly automated (two-person crew) tug has a bollard pull of 58 tons. Several ports have built or are building new tugs in the 55–70-ton bollard pull range.

Safety and biosecurity

The 1989 local government shake-up abolished the harbour boards and transferred their commercial operations to publicly owned port companies (some of which were later partly privatised). Regional councils now regulate their waterways and protect the harbour environment. The gold braid, spit and polish and the heavy staffing of the old harbour departments became a memory.

The old Marine Department (latterly the Marine Division of the Ministry of Transport, then the Maritime Transport Division of the Ministry) was downsized in the government reforms of the late 1980s/early 1990s, nearly 20 years after automation had removed the last lighthouse keepers. A Ministry of Transport survives, a small agency responsible for policy advice and as the minister's 'agent for managing the interface with the transport Crown entities'. The main maritime bodies are the Transport Accident Investigation Commission (TAIC) and Maritime New Zealand. TAIC was formed in 1990 and is a standing commission of inquiry, which investigates marine, rail and air 'occurrences' (accidents and incidents). According to its website 'a standing Commission eliminates the time taken to set up individual judicial inquiries into significant accidents and retains the skills and experience necessary to undertake what can be very complex and far-reaching investigations into almost every facet of a transport system'. The Commission believes that its ability 'to carry out much of its investigation work in camera contributes to shorter and more effective inquiries because the parties assisting the Commission with its investigation can concentrate on the inquiry's purpose without the distractions, pressures, and posturing sometimes associated with public inquiries'.[11]

Maritime New Zealand (formerly The Maritime Safety Authority) was created in 1993. Its responsibilities include maritime safety, security and marine environment protection; amongst other things, it manages the lighthouse system and pollution responses. It has been very active, especially against old or unsafe ships. Most of the defective rust buckets have been old Russian or Ukrainian trawlers and supply ships, some of which have cluttered ports for years, their crews in some cases reduced to begging.[12] In 2005 Maritime New Zealand took global giant MSC (Mediterranean Shipping Company) to task for shunting 25–30-year-old converted container ships down to this part of the world. Speaking about the deficiencies of some of these elderly ships, an official warned that 'substandard ships will not be tolerated'.[13]

Part of the concern about elderly ships — single-skinned tankers especially — is environmental. In fact, environmental considerations are now secondary only to human safety when a ship gets into trouble. Almost as soon as the crew is helicoptered away to shore, officials start removing fuel and lubricants. A typical recent example was the loss of the 456-ton fishing vessel *Seafresh 1* off the Chatham Islands in 2000. In the early hours of 9 March fire broke out in the ship's switchboard. It was quickly extinguished, but when water began leaking into the engine room, the master accepted a tow to Hanson Bay on the eastern side of the island. The company dispatched another trawler from Wellington to tow back the *Seafresh 1*, but on 17 March the vessel suddenly developed a list and sank by the stern off Owenga in Hanson Bay.

OPPOSITE, TOP By the twentieth century just one or two big ships a year were being lost. The Union Company's trans-Tasman freighter *Waikouaiti* (3926 tons) was wrecked on Dog Island near Bluff on 28 November 1939, fortunately without casualties.
Gift of Wellington Museums Trust, New Zealand Maritime Museum (2012.0.1982)

BELOW Regulators discourage visits by inadequately maintained or overage ships such as the Mediterranean Shipping Company's 36-year-old *MSC Sariska*, seen here at Lyttelton in November 2006. Earlier, Maritime New Zealand had required MSC to repair its elderly ships.
Gavin McLean Collection

> **Everyone knows that even with every available technical aid, the prospect for an environmental disaster is just a misheard helm command or a machinery failure away.**

Thoughts now turned from salvage to environmental protection: the *Seafresh 1* had 85 tonnes of diesel fuel aboard and another 10 tonnes of hydraulic and lubricating oils, all a potential threat to the island's rare seabirds. Oil response equipment and personnel were sent to the island while MSA and DOC officials put a boom across the Hikurangi Channel and divers sealed the ship's tanks. Then a new threat was discovered: the invasive Asian seaweed undaria on the *Seafresh 1*'s hull. On 25 March the Ministry of Fisheries declared the trawler a biosecurity risk and gave Seafresh New Zealand Ltd 30 days to remove it from the Chathams.

Easier said than done. Within days the underwriters had paid out and United Salvage Ltd had dispatched the *Southern Salvor* to New Zealand. It arrived off Owenga on 26 April, but although the masts and gantries were cut from the wreck and most of the oil was pumped out, the wreck's position and the winter weather frustrated United's plans to refloat the hull upside down and tow it into deep water for scuttling. So, after establishing a safe depth over the wreck by removing the rest of the masts, and scraping the hull to remove the undaria, the salvors sailed away on 21 June, nearly three months after the trawler first hit trouble.[14]

Ships still get into trouble in the twenty-first century. In 2002 two large bulk carriers, the *Jody F. Millennium* (6 February) and the *Tai Ping* (7 October), ran aground at Gisborne and Bluff, respectively, each requiring expensive salvage. The following year two large tankers, each capable of carrying over 100,000 tonnes of petroleum products, the *Capella* Voyager (16 April) and *Eastern Honor* (27 June 2003), scraped the bottom at Marsden Point. Even more alarmingly, that year the iron sands carrier *Taharoa Express* shut down engines 3.6 nautical miles off the coast west of Dargaville and drifted all night after cracks were discovered in its propeller shaft. While tugs sped from New Plymouth and Auckland, the MSA declared a Tier 3 oil spill response. Too deep to enter any New Zealand port for repair, the ship (which was carrying 130,000 tonnes of iron sands) was towed by the *Sea Tow 25* and *Tuakana* well out into the Tasman where the tug *Kayo Maru* picked up the ship to tow it back to Japan.[15] Months later the ship was in trouble again. On 22 February 2004 the *Taharoa Express'* engine failed in heavy seas while attempting to connect to the single point mooring buoy off Taharoa. Caught by the heavy swell, the giant ship began drifting towards the shore and was brought up only by dropping the anchors. Eventually the engine was restarted.

Speaking at a conference in 2003, Greg Cox from Discovery Marine warned that the combination of increasing ship size and ever tighter voyage scheduling was placing greater pressure on ship operators and port companies to push the limits.[16] By 2005 Maritime New Zealand had placed an oil skimming barge at Marsden Point, and there was talk of the need for a salvage tug for the country, but everyone knows that even with every available technical aid, the prospect for an environmental disaster is just a misheard helm command or a machinery failure away.

Helicopters, salvage vessels and mobile cranes make it easier to salvage vessels, such as the trawler *Southpac*, seen here at Castlepoint on Boxing Day 1991.
Michael Pryce

2

Drunk, deranged or dangerous — the human factor

Sometimes the misdeeds of individuals found their way into even the driest of reports. 'Two [lighthouse] keepers have been dismissed during the last two years,' Secretary of Customs William Seed wrote in the sixth annual report of the Marine Department in 1871: 'one for intemperance, and the other because it was found that he was constantly quarrelling with his assistants, and that he had such an ungovernable temper as quite prevented anyone from being able to remain with him.'[1]

If civil servants could be so uncivil, we should not be surprised to find others behaving badly. Nineteenth-century seafaring was such a nasty, brutish life that it is surprising how few ships were lost because of booze, brutality or insurrection. Even so, human failings squandered many lives and much valuable property. Sometimes the rot started at the top, on that holy of holies, the poop deck. There were numerous cases of 'hard-driving' masters pushing their crews to near-mutiny and more than a few of booze-befuddled masters being relieved of command by their subordinates.

Our window on this past is patchy, since magisterial courts of inquiry tended to concern themselves only with the behaviour of the master and his officers. The ordinary seafarer was usually forgotten, by his employer as well as by the court, for he was a mere wage worker. Men signed on a ship at the start of a voyage and were paid off at the end, however that end came about. In colonial times, a shipwreck could be costly for the seafarer. Until well into last century, shipping companies seldom compensated them for the loss of their personal effects in a wreck; in fact, shipowners invariably told their local agents to lay off the men as soon as the ship was abandoned to the underwriters. It was common for men to find themselves in a foreign country with only the clothes they stood in.

Officers and gentlemen?

One particularly sensational incident took place on 27 April 1864 aboard the migrant ship *Flying Foam,* bound from England to Auckland with almost 150 passengers. The trouble began when

OPPOSITE New Zealand's worst wreck occurred on 7 February 1863 when the steam corvette HMS *Orpheus* struck the Manukau bar while entering port. One hundred and eighty-nine of the 259 aboard died.
Auckland Libraries Heritage Collections, 7-C6

Huddart Parker's trans-Tasman liner *Ulimaroa* ran aground on Three Kings Island on 9 December 1902 while carrying 136 passengers and a crew of 58. Everyone got into the boats but eight subsequently died aboard one raft, and another boatload was never seen again. In all, the lives of 28 passengers and 17 crew were lost in the aftermath of the sinking. Most survivors reached Auckland wearing only the scantiest of clothing.

This wreck sparked the formation of the country's first shipwreck relief society. Until then communities had aided wreck victims and their dependents in an ad hoc manner, throwing concerts, auctioning goods or appealing through newspapers or shop window advertisements for money and clothing. Permanent shipwreck relief societies existed in Britain and in New South Wales, so in 1902 J.A. Park, the mayor of Dunedin, assisted by John Kerby, a young clerk familiar with the NSW organisation, formed the Shipwreck Relief Society of New Zealand to assist victims of shipwrecks and to recognise acts of bravery during wrecks. The broad-ranging committee included businessmen, political conservatives and labour union leaders.

The society changed its name to the New Zealand Shipwreck Welfare Trust in 1998. Still run from Dunedin, it continues to aid families of maritime misfortune, these days mainly commercial fishers.

a crewman assaulted a second-class passenger. The third officer attempted to apprehend the culprit but was forced back by the man's friends. The master asked the men to consider submitting 'to proper authority' and ordered the first-class passengers to arm themselves.

This done, and the decks cleared of women and children, he formally requested the crew to hand over the accused. Between them the chief, second and third mates and the surgeon carried the kicking, screaming, swearing man aft. One seaman threatened them with a knife, but half-heartedly, and he did not stop them from securing the alleged offender.

From then until 13 July, when it reached port at Auckland, the *Flying Foam* was a ship divided. On 10 July, with their destination in sight, seamen refused to start bringing up passengers' luggage. They were locked up after several warnings but then five from the next watch struck in sympathy with their colleagues. Ordered to temporary imprisonment in the forecastle, several ostentatiously sharpened their knives on the grindstone as they walked forward. Once there, they amused themselves by smashing fittings and swearing loudly. The carpenter erected a set of stocks, but the seamen wrecked them and freed the men they were holding.

The *Flying Foam* was now largely under the control of the officers and passengers, none of whom had booked for a working holiday. Their inexperience probably added a fortnight to the voyage, but they got the ship to Auckland, where the armed constabulary dragged 16 sailors off to Mount Eden prison.[2]

The legal system loaded the dice against the ordinary sailor. Only from 1845, five years *after* the signing of the Treaty of Waitangi, did English law provide for the *voluntary* examination of a ship's master. Until then, anyone with the right connections or the gift of the gab could take command of a ship, regardless of whether he could read a chart. With the passing of the 1854 Merchant Shipping Act, a hefty 548-clause

WRECK OF HMS *Orpheus*

In terms of lives lost, New Zealand's costliest wreck was that of HMS *Orpheus* on 7 February 1863 in the Manukau Harbour. Of the 259 officers and men aboard, 189 died, including Commodore William Farquarson Burnett, C.B.

HMS *Orpheus* was a modern steam corvette. Launched in 1860, the ship had commissioned at Portsmouth the following year and was serving in Australia as station flagship. The 1706-ton vessel carried 20 broadside guns and one Armstrong 110-pounder mounted on a slide. Although fully ship-rigged on three masts, the *Orpheus* was a screw steamer. The Humphrys & Tennant steam engines gave a service speed of about 10 knots.

HMS *Orpheus* came to New Zealand from Sydney to deliver stores for Royal Navy ships stationed in New Zealand. Commodore Burnett had planned to enter the Waitemata. Unfortunately, contrary winds in the Tasman during the first part of the voyage delayed the ship, so Burnett, wanting to berth by the 7th, headed for Onehunga. The Manukau Harbour is large and shallow, an intricate network of mud and sand banks intersected by numerous channels. These channels change rapidly, and the bar at the entrance, as mariners knew, is dangerous. The harbour had claimed several ships since the wreck of the barque *Orwell* in 1848 — some with all hands.

When the *Orpheus* arrived off the Manukau bar at about 1 p.m. on 7 February, the weather was fine and clear, and the breeze moderate. After being signalled to 'take the bar', Burnett entered, following signals from the shore station. Unfortunately, recent changes to the main channel meant that the charts carried on the *Orpheus* had to be read in conjunction with more up-to-date sailing directions, which the ship did not have. The sandbar had grown and extended out into the old channel. Furthermore, only two men aboard (one a prisoner who, extraordinarily, is said to have remonstrated with Burnett) had sailed on the Manukau Harbour. The *Orpheus* crossed the bar safely but struck at the northern edge of the submerged Middle Bank at about 1.30 p.m. The ship broached to, putting it in serious danger, because the combination of sea against tide and a strengthening sea breeze had by now built up the wave height. The shallowing seabed turned the waves into destructive breakers, which swept the decks, forcing men into the rigging and bumping the hull, flooding compartments.

The ship's predicament went unnoticed by signal station staff or other shipping, so Commodore Burnett dispatched boats to seek help. By the time the steamer *Wonga Wonga* reached the *Orpheus* at 6 p.m., the ship was almost completely submerged, and the crew were clinging to the rigging. The *Wonga Wonga* released lifeboats to drift near the *Orpheus*' bowsprit and jib boom, where the water was deeper. Lieutenant Hill and Maori boatmen rescued some of the men who jumped.

The *Wonga Wonga*, unable to get close, stood by while the seas worsened. The *Orpheus*' heavy guns broke loose, adding to the damage. At 6.30 Burnett urged the men to say their prayers and to look after themselves. 'I shall be the last. The Lord have mercy on us all.'

Around about 8.30 p.m. the masts collapsed one by one taking the majority of the officers and men with them. Ironically, by daylight the wind had subsided, and the sea was perfectly calm. All that could be seen of the *Orpheus* was the stump of one mast and a few bare ribs.

Although Aucklanders contributed generously to a welfare fund for survivors, the aftermath of the wreck was clouded by controversy over responsibility for the accident. The loss was caused — like so many — by a series of minor mistakes, some from the signal staff, others by the ship's crew, and the delay in appreciating the ship's plight undoubtedly cost many more lives than it should have, but the principal reason was that Commodore Burnett was in a hurry. Had he gone to Auckland, had he sent in a boat to sound the channel, had he requested a pilot, had he listened to the ship's sailing master or the deserter ... all might have been very different.[3]

document, the rights of employers and employees and matters of safety were finally defined in reasonable terms. A year later, following tragedies aboard migrant ships such as the New Zealand-bound *Lloyds,* which lost 65 of its 211 passengers because of poor provisioning and the incompetence of both the master and the ship's surgeon, the British Parliament tightened up laws with the passing of the 1855 Passengers Act.

Even so, justice in New Zealand was amateurish. Otago's first resident magistrate A.R.C. Strode, had no legal training, and the justices of the peace who helped him were little better, reflecting upper-middle-class prejudices in their sentencing. Elsie Locke, in an unpublished study of early Otago seafaring, records that the magistrates tended to exercise the maximum penalty of 12 weeks' gaol, regardless of circumstances.[4] Since labour was usually in short supply in the young colony, sentencing men to hard labour benefited such employers as these very same justices of the peace. On occasion masters even tried to detain men they only suspected might desert. In February 1854 Captain Ross of the barque *Clutha* tried unsuccessfully to have two hospitalised seamen detained until the ship's sailing date.[5]

When nine striking seamen were imprisoned that same year after a protracted 140-day voyage aboard the *Thetis,* calling the old craft unseaworthy, a correspondent to the *Otago Witness* exclaimed in vain: 'I understand that they are to be put in irons, and sent on board that leaky old tub the *Thetis,* which would have sunk with her valuable cargo in the Indian Ocean but for the activity and energy of these hardy tars who … will not cheerfully remain and pump out this old sieve till she drifts to England at the rate of 5 knots.'[6]

Provincial superintendent Captain William Cargill disagreed. Responding to a plea for clemency and for extra rations ('Monson's Hotel', as the cramped little gaol was called, was not known for the liberality of its fare), he backed the captain, observing that 'any mitigation of the punishment would be an encouragement to other crews to do the same'.[7]

Such blatant class bias sometimes provoked recent migrants to feed and house seamen taking 'French leave' from colonial gaols. The passengers, fresh from close observation of shipmasters and officers, were sometimes more tolerant than the judiciary, which was biased towards its own mercantile class. When the *Strathallan* arrived in Otago in 1858, after a voyage in which Captain John Todd spent most of his time drunk, and had been relieved by the chief mate, the entire crew refused duty but was convicted of disobedience. When they tried to prosecute Todd for supplying inadequate provisions, the same judiciary threw out the case.

Another 'Strath boat', the *Strathfieldsaye,* brought more tales of woe into port a few months later, with one settler's diary alleging that the master and purser had conspired to swindle the passengers out of their rations (a frequent complaint). Just before entering the heads, Captain James Brown was heard to threaten to shoot any passengers or crew who complained about him. He then swung around a signalling cannon on the poop, aimed and fired, slightly injuring a passenger in the hand. The captain, who had taken up with what some migrants called 'a cheap whore', was later fined for serious breaches of the Passengers Act. Colonist Daniel Brown recorded bitterly that while 17 *Strathfieldsaye* sailors went to prison for opposing this maniac, Brown sailed back to Britain aboard the *Strathallan,* having paid only light fines.

Drink was implicated in the destruction of a number of ships, as other southern examples showed. Visitors to the beautiful Otago Peninsula can still see on Victory Beach in Wickliffe Bay the rusting remains of the steamer of that name that went ashore there on 3 July 1861. The 737-ton ship had been the toast of the trans-Tasman passenger service since entering the run between Melbourne, Port Chalmers and Lyttelton earlier that year. After clearing Otago Heads, Captain Toogood had plotted a course to pass three miles clear of Cape

The boiler of the steamer *Victory*, photographed on Otago Peninsula in 1952, 90 years after going ashore there.
Ian Farquhar Collection

Saunders and had handed command to the third mate briefly until the chief mate took over. When the ship struck, Toogood found chief mate George Hand 'tipsy' and called the saloon passengers together to verify the man's condition. Fortunately, the ship had struck a soft, sandy beach, so no lives were at risk while Toogood covered his butt. Passengers, mail and cargo were landed, and a week later Dunedin was treated to Toogood's prosecution of George Hand under the Merchant Shipping Act 1854.

The second mate, the chief engineer and others testified to Hand's intoxication and his absence from the bridge when the ship struck, and he was sentenced to three months' hard labour in the Dunedin gaol. Toogood got vengeance but could not entirely save his reputation. The court, while sentencing George Hand, observed that even if the officer had been sober, he may not have been able to save the *Victory*. The compass had been inaccurate, and the evidence presented suggested that discipline aboard the ship had been lax.[8]

For a while the *Victory* looked as though it might cheat death. The ship was beached far above the shoreline and had sustained little damage. A year later a Sydney engineer began salvage work, and by November the *Victory* was afloat and steam was being raised. Then disaster struck. At a crucial moment, the ship's anchor chain parted and it drifted back on to the beach, where it later broke up.

An even more sodden story took place off the South Otago coast a decade later when the immigrant ship *Surat* was wrecked near the entrance to the Catlins River on New Year's Day 1874.[9] Its master, Captain Edmund Johnson, was unfamiliar with the coast and struck Chasland's Mistake on the evening of 31 December 1873, bumping several times. There was a brief panic before the crew ordered the passengers below.[10] Johnson took the *Surat* out to sea and ordered the passengers — women as well as men — to join the crew on the pumps. But only the port side pump and a small fire-engine pump were working, so the water slowly gained ground.

As exhaustion sapped the energies of the people on the pumps, Johnson unwisely ordered his steward to pass around gin 'restoratives'. Things quickly got out of hand. The first mate was soon so drunk that he even tried to stop the pumps and by the time the coastal steamer *Wanganui* pulled abreast of the *Surat* in the early hours of 1 January, Captain Johnson was also showing signs of inebriation: he refused to signal for help and when a frightened passenger begged him to do so for the sake of the children aboard the ship, Johnson threatened to shoot anyone who tried to communicate with the *Wanganui*.

More scenes of anarchy took place on the *Surat*'s sinking deck that morning as officers, men and passengers struggled amongst themselves. Eventually, Johnson agreed to take the *Surat* into (Bloody) Jacks Bay, south of the Catlins River mouth, to enable the passengers to get ashore in the boats. He would then take the *Surat* to Port Chalmers for repairs. He accomplished the former, landing them in the sparsely settled sawmilling community, but the *Surat* travelled only a short distance to the northern side of the Catlins River mouth, where he beached it on what is now known as Surat Beach. The passengers were rescued by the French warship *Vire,* then lying at Port Chalmers, and the *Surat* was left to break up. Johnson ultimately had his certificate cancelled and in a separate criminal action was gaoled for two months.

That second legal penalty ignited a bitter debate. On 22 January ship masters at Port Chalmers took up a collection for Johnson and began a subscription (£52), criticizing reliance on compasses in iron ships, blaming the strong set on the southeast coast and sending a petition to the Board of Trade in London. The wreck's historian considers the verdicts 'harsh' and that the brouhaha may have been intensified by the clash of values between passengers — 'upright, teetotal, good Christian folk' — and the rougher seafarers, but it is hard to disagree with the courts' verdicts or with those newspaper correspondents or editorial writers who took comfort from courts punishing 'reckless negligence'.[11]

The Union Steam Ship Company's 'Instructions to Officers' of 1878, the start of the codification of practices and working conditions that improved working conditions in the bigger companies.
Union Shipping Group

Discipline

Gradually, though, the bigger lines weeded out miscreants and imposed higher standards. The reasons are obvious. Although sailing ships followed the dictates of time and tide, steamships were so expensive they had to be worked

INSTRUCTIONS TO MASTERS

For decades 'the red book', as the *General Instructions for Captains and Officers* was known, regulated conduct aboard the Union Company's ships.

The 1878 edition was short and sweet — a single page sufficed. When the 1895 edition was distributed, masters were instructed to sign for the copies issued (to go to the captain, chief, second and third officers and the chief engineer) and to return every copy of the superseded 1886 edition.

The *Instructions* specified everything from hours of work, to clothing requirements and emergency drills. Officers' coats, for example, were to be 'a double-breasted sac of sufficient length to entirely cover the sea, to have five of the Company's buttons on each side, four being visible when the collar is turned down, and three vest buttons on the side of each cuff'. Seamen were to wear 'navy blue trousers, jerseys and bonnets of the Company's pattern' which could be bought 'at cost price from the Purser on board'.

The booklet set out the officers' duties. Captains were to be between the ages of 25 and 40 and to have served at least nine years at sea. The age range for chief officers was 24–35 and they needed to possess a master's certificate; they were told to procure pilotage exemption certificates for all ports 'without delay', the fees incurred being refunded when they were promoted to captain. At the bottom end of the hierarchy, fourth officers were to be 20–28, to have served at sea for five years and have a second officer's certificate.

The captain had overall responsibility for the running of the ship. The chief officer had 'general superintendence of all work on board', the second officer looked after the boats, life-saving equipment and the state of the holds, the third the mails, lights and signal guns and the fourth (where carried) passengers' luggage and the custody of the luggage book.

A large section of the *Instructions* set out the requirements for documentation — the night order book, bridge book, azimuth book,[12] observations and the maintenance of navigational instruments.

Officers were instructed to observe every care when approaching land. In foggy conditions they were to stop the ship if they had 'any doubt regarding the safety of the course the steamer is following'. If senior officers considered a course the master was taking dangerous they were 'to record the fact in writing' and send it to the company's marine superintendent.

Masters were 'strictly prohibited from racing their ships with other steamers'. Fire and boat drills were to be conducted once on every trans-Tasman round voyage and at least monthly on other ships. All boats were to be maintained carefully. Ships' holds were to be sounded night and morning, at sea and in harbour.

The company was strict about alcohol. Passengers and crew were not allowed to bring private supplies aboard and officers were not to drink on sailing day or at sea. 'When promotions are being arranged, preference will be given, other considerations being equal, to those Captains and Officers who are total abstainers.' To make sure directors knew what was going on, all liquor supplied to officers had to be signed for and settled monthly through the purser's accounts — cash payments were forbidden. Any drunkenness by officers must be reported to the managing director 'when the offenders will be summarily dismissed from the service'. Cardplaying was also banned and officers were expected not to use 'coarse language' and to make provision for church services on Sunday where practical.

---

efficiently to get a satisfactory return on the shipowners' investment, with this giving a new emphasis to training and management.

The Union Company led the way in imposing greater order on officers and men. Whereas most owners were usually content to rely on the sanctions imposed by courts of inquiry, the Union Company took a tougher line. The company's *Instructions to Masters,* commonly known as the *Red Book,* set out its expectations regarding dress, drills and behaviour. It used carrots as well as sticks. From 1878 every master who completed a 12-month, accident-free term received the bonus of a month's salary, but the unlucky, even if blameless, were dealt with harshly, regardless of rank or length of service,

> **Seafaring was a hard life that drove many to the bottle. The hostels clustered around the port areas of most towns sold ferocious concoctions.**

as the case of Captain M. Carey showed. In 1883 this senior master ran the crack trans-Tasman flyer *Rotomahana* aground on Waipapa Reef. It was recovered, and the court absolved him of all blame, but the directors, probably still smarting from the £4255 repair bill, disciplined Carey:

> The Directors have considered all the circumstances surrounding the recent stranding of the *Rotomahana* while under your command and have come to the conclusion that although the magisterial enquiry exonerated all those on board from blame, on general grounds the responsibility must rest entirely with you — the Master being the officer entrusted with the care of the ship and the lives on board.
>
> In the present case you had a fine night, reliable compasses and a good chart and with these such an accident should have been impossible, while a cast of the lead would have determined your accurate distance from the shore at any moment. There can be little doubt that the disaster was brought about by the steamer's distance from Slope Point at three o'clock having been overestimated and by insufficient allowance having been made subsequently for the inset of the current and probably by a little careless steering.
>
> The Directors are therefore forced to the conclusion that you cannot be held altogether free from responsibility for the accident; having regard, however, for your previous good judgment and discipline displayed on board while the steamer was on the reef — and the fact that there has heretofore been no recognised rule in the Service for dealing with such cases, they have decided to allow your position in the service to remain undisturbed.[13]

But not untainted. A circular sent around the fleet outlined Carey's offence and warned that masters would not be reinstated 'unless the accident is proved to have occurred under such exceptional circumstances as to entirely exonerate him from blame'. Indeed, for decades, any master involved in a serious accident, even if acquitted by a court of inquiry, was likely to find himself marched down the gangway.

The coastal companies also battled the demon drink. Seafaring was a hard life that drove many to the bottle. The hostels clustered around the port areas of most towns sold ferocious concoctions. Firemen, performing a thankless, poorly paid job, were the worst offenders, often going on binges that sometimes prevented ships from sailing to schedule. This was especially common at the smaller ports, far from the prying eyes of their superiors. As Union Company official Robert Strang reported in March 1892, officers usually hit the bottle in the 'outports', being 'too knowing to indulge in any large port'.[14] In February and March of 1892 alone, the company's officers had reported Captain Downie of the *Wareatea* and the officers of the *Taieri* tippling, the first mate of the *Oingadee* being 'the worse for alcohol' and a seaman from the *Orowaiti* having 'got drunk and bit Captain Adams on the nose'.[15] In August 1900 the master of the coaster *Kini* tangled the Normanby Wharf chains while leaving Oamaru Harbour — the branch manager finding the *Kini*'s Captain 'in a hopeless state of collapse', after having unwisely mixed whisky with morphine.[16]

Sometimes the company gave errant staff a second chance if they signed the pledge. This did not always work: in November 1905 Strang reported that Captain McNair was drinking heavily, just eight months after signing the pledge under duress. He was sacked.

Every company kept a watchful eye on its seagoing staff. The archives of Napier-based Richardson & Co. contain this semi-literate passenger's complaint against Captain Thompson of the *Kahu*:

> I should like to bring to your notice the goings on on the *Kahu* the night he left Napier the master was silly drunk on the bridge at between 12 midnight and the time she landed her passengers he was so silly that at 4 o'clock in the morning that he had to call the Chief Engineer out to tell him where he was, also the language (filthy) he was using on the bridge where there were 2 female passengers and 2 children laying I should like you to warn him if only for the sake of the passengers that travel by your vessels and the safety of both the crew and the passengers and vessel for on Sunday morning her did not know here he was as we were inside of the light instead of outside.[17]

Richardsons fired Thompson, who had already received several formal warnings. It had no choice. With lives, a valuable ship and cargo at stake, such behaviour could not be tolerated.

Drunkenness and indiscipline lingered on into the early part of last century, especially with the smaller lines, but by the start of the inter-war period improved training and higher entry standards had weeded out most of the less capable officers and men, and the conflict between the two groupings was, to a greater extent, being fought out through trade union action. Seamen and firemen still went on 'benders' but by the 1960s the problems had largely disappeared.

ABANDON SHIP!

As the next chapter will show, the West Coast bar river harbours exacted a heavy toll on port users.

The *Perth* was just one of many fine ships to leave its bones at Greymouth's notorious entrance. On the morning of 13 November 1921, the Melbourne Steamship Company's steamer was lining up on the entrance, following in the steamers *Ngakuta*, *Kaiapoi* and *Kamona*. Although a heavy sea was running, there were no grounds for undue concern. The *Perth* had been a regular caller at Greymouth and was bringing in a cargo of timber and was due to load the usual return freight of coal. The vessel had been built at Sunderland in 1897 and was a steel, screw steamer of 1799 tons gross.

The bar had been sounded a couple of hours earlier, so Captain MacDonald was surprised when his ship bumped just outside the North Tiphead. The ship recovered from that blow and from a subsequent one, but a third, much heavier blow caused the *Perth* to ground heavily and to lose all way. It was now out of control. The waves quickly forced the *Perth* stern around, leaving the ship lying 50 metres off the tiphead, broadside on to a south-westerly sea.

The tug *Westland* was beaten back by the heavy breakers, which made it impossible to render assistance. Seeing that the situation was hopeless, Captain MacDonald, ordered the crew to launch the lifeboats. They had a perilous run through the breakers but within two hours of the stranding everyone was safely ashore.

The *Perth*, however, soon broke up. The court of inquiry exonerated MacDonald, finding that the ship had hit the seabed after three unusually large seas struck the ship, forcing the keel down on to the seabed. His certificate was returned, and he was praised for his skilful work in getting the crew ashore.

J.C.W. Fleming

3
Bar harbours — river ports

Most of New Zealand's mountain-fed rivers are short, shallow and dangerous. Few, apart from the Clutha, Whanganui and the Waikato, are navigable for any distance. For the nineteenth-century mariner, there were, if not a taniwha around every bend, enough natural hazards. They began at the entrance where dangerous bars — submerged ridges of sand that shifted with storms — guarded river mouths and provided an unpleasant obstacle to ships entering rivers. Nor were these river mouths static. Many moved with the seasons, cutting out fresh entrances whenever storms savaged the coast.

Once inside the river, the mariner faced new dangers. Freshes could scour out banks or bring down debris and snags, fouling propellers and puncturing hulls. Sometimes floods tore ships from their moorings. Yet, despite that, river ports were preferable to the horrors of open roadsteads (see chapter 4), and communities made the best of any little waterway. In 1868, for example, settlers at Kakanui, North Otago, started deepening parts of their river, which, the *Otago Witness* said, 'at some places did not carry much more water than would be required for floating a duck'. Anything was better than nothing.

Most communities did their best to improve upon the parsimony of nature. They surveyed rivers, marked dangerous snags and rocks, and selected wharf sites. First to go up were signal stations and simple wooden jetties, which often sufficed for the next decade — all that were needed before road and rail links killed their trade. Later, the survivors might engage an engineer such as Englishman Sir John Goode to prepare expensive plans to handle bigger ships, usually by straightening up the entrance channel and pushing it out to deeper water between the protective arms of breakwaters or training walls.

Hokitika was the most notorious river port. Although ill-equipped by nature for such a role, Hokitika gained unofficial port status in 1864 after miners discovered gold in the hinterland. That December Captain Leach took the small steamer *Nelson* across the bar and into the shallow,

OPPOSITE The collier *Kaponga* (2436 tons) was one of the larger ships lost at Greymouth. After touching the bar while leaving port on a falling tide on 28 May 1932, the ship became helpless after damaging its steering gear. The tug *Westland* failed to swing the *Kaponga*'s bow round to face the sea, leaving the freighter broadside on to the waves. Within a day all hopes of salvage had gone, and large rollers were pushing the wreck towards the tiphead.
Gavin McLean Collection

snag-infested river. Leach got away with it, but others would not.

Things began optimistically. The Canterbury authorities, delighted to compete with Otago's goldfields, gazetted Hokitika as a port in March 1865 and started building wharves, setting aside reserves and knocking together a customs house. By the following year a tolerably good wharf, Gibson Quay, was cluttered with ships, many direct from Australia. Vessels too big to cross the bar (which shifted constantly and could vary in depth from less than a metre to over two metres) lay off the entrance, discharging passengers and goods into smaller craft.

But it was a curse of a place. The holding ground off the bar was poor. Ships forced to anchor there were at the mercy of any storm that blew in from the Tasman. Neither buoys nor surf-boats could be used. Then there was the entrance — narrow,

ABOVE The *Jane Douglas* (95 tons, 1875) alongside Gibson Quay, Hokitika. This tough little ship had several scrapes there between 1902 and 1907. She was lost between Stephens and D'Urville islands on 10 January 1912.
Ref: 1/1-002170-G, Alexander Turnbull Library, Wellington, NZ

OPPOSITE ABOVE & BELOW Few ports inspired more dread than Hokitika, the 1860s gold rushes' shipwreck capital. These photographs show seven ships ashore in 1865, Hokitika's worst year. The lower picture shows three ships: the three-masted schooner (right) may be the London-owned Sir Francis Drake, which struck the bar and then hit the beach while trying to enter port on 29 June 1865.
ABOVE Ref: 1/2-018448-F **BELOW** Ref: 1/2-050050-F, Alexander Turnbull Library, Wellington, NZ

shallow and unpredictable even at the best of times. It usually ran to seaward in a north-south direction, forcing vessels entering and leaving to do so broadside on to the prevailing seas. Even inside the river, skippers had to remain wary, because floods often swept down the river, diverting the channel away from the wharf, effectively closing the port for weeks. Sometimes the torrent carried ships down the river and out over the bar.

Paddle tugs ferried passengers and cargo from ships in the roadstead and helped small craft in and out of the port. In August 1865 McMeckan Blackwood & Co. of Melbourne set up the first service, charging outrageous fees for the *Yarra*. Fees fell after Dunedin merchants muscled in on the business but, even so, Hokitika was never cheap to work.

The price was paid in lives as well as cash. Between 1865 and 1867 there were 108 strandings — one every 10 days! — 32 of them total losses. The causes varied: some ships grounded going out, others coming in, some snapped their moorings during freshes, sailing craft lost the wind at a critical point while crossing the bar and, on rare occasions, steamers had their fires doused by heavy seas. All ended up on the bar, some to be refloated and sail again,

The steamer *Torgauten*, stuck fast on the Hokitika bar in August 1904, after grounding owing to what the *Grey River Argus* newspaper described as the 'narrow and tortuous' nature of the channel. A month later she grounded again in Golden Bay — ironically while bound for Hokitika with a cargo of dredging equipment. Ultimately such maintenance would be futile, and by the time Hokitika closed to shipping in 1950 the port had become badly silted.

West Coast New Zealand History, the steamer *Torgauten*, stuck fast on the Hokitika bar, 1904, https://westcoast.recollect.co.nz/nodes/view/25048

Greymouth was safer than Hokitika but like all bar harbours, had to be treated with respect. On 22 December 1884, when entering the river against harbour signals, the *Star of the South* was caught by the current and flung around. Captain Charles Hodge hoisted the auxiliary sails and tried to sail upriver, but the current was too strong, and the *Star* struck near the entrance. In January 1885 a flood swept away the remains.
Hocken Collections, Uare Taoka o Hākena, University of Otago. Star of the Sea, 1884. James Ring photograph, Box-063-002

others to provide firewood for the townsfolk.

'From the distance to the river where the *Montezuma* has been cast high and dry on the sands, the picture is one that cannot be equalled in the colony and perhaps not in the world,' the local rag observed one day. 'In one spot the last remnants of the *Oak* may be observed — showing even now how well and faithfully she must have been built; further on a confused mass of ruin, a heap of splintered planks and ribs, marks the place where the *Sir Francis Drake* and the *Rosella* finally succumbed to the force of the waves. Yonder can be seen the mast of the *Titania,* and nearer home, what is left of the steamship *New Zealand*.'[1]

In *Hokitika: Goldfields Capital,* Philip Ross May records that Hokitika produced 'beach rakers' — men who lived off scavenging for material washed from wrecks. They worked in pairs, usually by moonlight, using hook and line to drag in anything of value. Wrecks kept many locals in food and drink. 'First come the auctioneers, who try to assume a serious aspect (with about the same results as mutes returning from a funeral) as they say aloud, 'sad thing, another wreck, I'm afraid'; while inwardly they chuckle to themselves, "Fi fo fum, I smell another sale",' Ross recorded. 'They are closely followed by the shipping agents, who affect big sou'-westers and oil-skin coats, and who in virtue of their business are very nautical ashore, and are looked up to as authorities; then come the mass, the consignees, the draymen for jobs with perhaps Thatcher [a goldfields balladeer] to get some fresh material for 'a screaming new local'.[2]

Space precludes listing every Hokitika wreck here but let us take one day — 9 May 1867 — to see how easily a single storm could endanger shipping inside and outside the port. The first casualty was the brigantine *Goldseeker,* which had arrived off the bar on the evening of 7 May. An old (1852) craft bought recently by Hokitika merchants Spence Brothers, and sporting an appropriate name, the *Goldseeker* carried 200 sheep and general cargo. Since the weather looked threatening, Captain Wilkinson decided to haul offshore under shortened canvas and let the storm blow over. The next day, in fine weather, a steamer towed the *Goldseeker* back.

Unfortunately, they missed the tide by an hour. Wilkinson, left on his own again, dropped anchor for the night expecting to enter the river the next morning. During the night though, the weather again deteriorated, and he had to drop the second anchor. At first, they held but, as the westerly gained strength, the *Goldseeker,* its bow held down as it strained at its anchors, started shipping seas and filling with water. When the inflow topped the pumps' capacity, Wilkinson, in an action later criticised by the mate, cut the cables, hoisted sail and ran his ship up onto the beach to save lives. It soon broke up. While that was happening outside the entrance, the weather was also affecting shipping alongside the quay as the *West Coast Times* reported.

> Yesterday may be truly considered a black-letter day for the port, as it witnessed the destruction of one fine vessel outside, and whilst that was going on a scene occurred in the river that caused an hour or two's extreme excitement and the most lively apprehensions for the safety of every vessel moored at the Transit Shed, and that of the wharf also, which was at one time threatened with total destruction.
>
> Luckily the crews were stirring, having just turned out at about half-past six, otherwise the most serious results might have ensued. The heavy rains of the past night considerably increased the volume of the river, which at low water rushed down with great force and set full upon the *Bella Vista,* the headmost vessel of those below the shed. She was well moored head and stern, but by some means her bow took a shear off, and the current getting between her and the wharf brought such a strain upon her head lines that the mooring post, unable to bear it, snapped short off, and in an instant, she swung across the stream, and grounded bows-on to the shallow bar in mid river. Dropping astern at the same time, she fouled the *Anne Moore,* snapped that vessel's flying jib-boom, and otherwise damaged her headgear, but not very seriously. The ketch *Florence* was moored outside the *Bella Vista,* and by this sudden change in position found herself in some peril, but her crew moved themselves, smartly, ran lines on shore, and soon got their little vessel out of difficulty, although she also grounded for a time on the middle bar. Meanwhile the *Bella Vista* lay athwart the stream, her bow on the bar and her stern hanging by the quarter warp, which got under the *Anne Moore*'s bowsprit, and threatened to tear it from the gammoning. The river thus being obstructed, rushed and boiled furiously against the wharf and in less than five minutes worked underneath the piles and washed away tons of the gravel used as the filling-in, and proceeded in its work of destruction with such rapidity that had not the deputy Harbour Master, dreading some mishap and hence on alert all night, given orders to cut the *Bella Vista*'s quarter-line, we feel certain that forty or fifty yards of the lower wharf would most likely have gone. The remainder would have received such serious damage that the whole thing would have needed to be rebuilt. The barque thus released swung to the stream and grounded fore-and-aft, permitting the river to flow onwards in its usual course, and thus saved the wharf.[3]

By the 1870s most of the gold was gone and Hokitika was exporting timber and other mundane goods. Sir John Goode's modern harbour works were finished by the end of the decade, but they almost finished the harbour board, which spent the rest of its long existence presiding over a diminishing trade. Hokitika was declared closed to shipping in 1950, to the relief of local administrators and the longsuffering Marine Department.

Coal port casualties

The West Coast's maritime trade settled down to coal and timber exports through Greymouth and Westport. Yet, even though ships grew bigger and better, the rivers, particularly the Grey, continued to exact their toll.

The Union Company had very little luck with the name *Hawea:* the first, as we shall see in chapter 9, sank at New Plymouth under peculiar circumstances, and the second went ashore at Greymouth. The second *Hawea* was a utilitarian-looking cargo carrier of 1758 tons, built at Dumbarton in 1897. Cheap and simple to operate, the ship called at West Coast ports regularly, loading coal and timber for ports on both sides of the Tasman.

The *Hawea* was leaving Greymouth for Australian ports on the afternoon of 30 October 1908 when it slammed into a heavy roller while crossing the inner bar. The huge wave struck the port bow, dumping the ship onto the seabed and causing it to lose steerage way. As Captain J.W. Burgess tried to regain control, the steamer crashed into the northern tiphead, grounding amidships, and listing alarmingly as huge waves broke over it. Deck cargo broke loose, endangering the lives of the crew, who were relieved to see the waves calm shortly afterwards, enabling them to come ashore on a safety line later that afternoon.

With the men safe and sound, attention now turned to the ship and its cargo. Company officials

The Port of Greymouth claimed another scalp late on the night of 3 September 1917 when the Opouri Shipping Company's steamer *Opouri*'s steering gear failed just as the vessel was crossing the bar. In a few short minutes, the ship swung around and was carried on the North Spit breakwater, where it hit heavily, bounced off and then stuck hard and fast. The 17 crewmen got ashore safely and although a locomotive hauled the *Opouri* shorewards away from the rocks and on to a small sandy beach, the ship proved impossible to salvage.

The court of inquiry exonerated Captain E.A. Cox, his officers and crew, putting the blame on a defective parting-screw on the starboard steering-rod coming loose at a crucial time. The loss of the 570-ton ship, just six years old, was a blow to the depleted wartime coastal shipping industry.

Gift of Wellington Museums Trust, New Zealand Maritime Museum (2012.0.3526)

were deeply divided. From Dunedin, Marine Superintendent Captain Col McDonald followed Managing Director Charles Holdsworth's line that the ship should be salvaged only if it was more profitable than allowing it to break up and the company collect the insurance money. On the coast, J. Daniels, who had successfully salvaged the *Mapourika* a decade earlier (q.v.), was committed to salvage, although whether it was the ship intact or just the cargo he had yet to determine. The conflict spilled over into the press, with the *Grey River Argus* spitting that 'to manage a job such as the *Hawea* presented by means of telegrams from business men hundreds of miles apart would be like managing the South African campaign from Downing Street; and we make no doubt at all that the disappointment in each case was largely due to a like cause.'[4]

By 1 December work gangs were jettisoning the timber into the river for others to recover and store near the Cobden Bridge. Daniels, still committed to saving the ship, was fashioning a canvas cover to put under the vessel before refloating it. The weather was moderating, and it looked as if he would succeed. But the coast's weather can be fickle, and Dunedin got its wish when a storm sprang up and pulverised the *Hawea*.

Greymouth claimed many more victims. Big ships lost there after 1914 included the *Opouri* (570 tons, 3 September 1917), *Perth* (1799 tons, 13 November 1921), *Ngahere* (1090 tons, 12 May 1924), *Kaponga* (2346 tons, 28 May 1932) and *Abel Tasman* (2042 tons, 18 July 1936). In nearly every case they grounded on the northern side of the inner bar, worked across the channel and hit the north tiphead where the sea pounded them apart.

The big, engines-aft collier *Kaponga*, its holds bulging with more than 2900 tonnes of coal, left the berth at 1530 hours on 28 May 1932 and lined up on the entrance. The ship was following the smaller *Kalingo* (drawing 5.4 metres) and should have had plenty of clearance. The harbourmaster said there were 6.7 metres of water on the bar and the *Kaponga* was drawing 5.6 metres aft and just under 5.5 metres forward. Conditions were ideal, with good visibility and a very calm sea. What Captain W. A. Gray did not realise, though, was that the *Kalingo* had lightly touched the inner bar on the way out.

The *Kaponga* struck just opposite the signal station and drifted about two ships' lengths before halting. It was the gentlest of impacts, evidenced by a loss of headway rather than a bump or jar. The sea was moderate but the swell swung the bow around to the northward, straddling the ship across the inner bar, despite Cray's best efforts to so keep pointing into the waves. The *Kaponga* was now bumping up and down on a falling tide. Three quarters of an hour later the tug *Westland* got a line to the *Kaponga* but the rope broke. Then a loud crash signalled that the ship had hit rocks. Its rudder now hung broken and useless, leaving the ship helpless.

Hopes rose briefly when the small coaster *Titoki* crossed the inner bar. Its crew threw the *Kaponga* another line, but that rope also parted and after standing by for a while, the *Titoki* continued on to Westport. By then the *Westland* had picked up a line from the *Kaponga* and was maintaining a strain to keep the *Kaponga*'s bow off the rocks on the northern tiphead, just 15 metres away. Perhaps it could hold the line until the big salvage tug *Terawhiti* arrived from Wellington. The harbour board wanted a quick fix — the *Kaponga* had blocked the port, trapping

Westport also claimed many victims. On 10 July 1888 the Union Company's *Suva* (293 tons, 1877) was washed onto the breakwater while leaving port and then carried south onto the beach. There the ship broke up within weeks.
Gift of Wellington Museums Trust, New Zealand Maritime Museum (2012.0.1875)

the *Komata* and *Karepo* and causing four ships to divert to Westport.

There was still hope. The *Kaponga* had taken on some water after grounding, but the pumps were keeping it under control. The crew decided to spend the night ashore and with the seas moderating, felt confident of refloating the ship. Flooding was limited to the fore- and afterpeaks, and the engine room remained dry.

But on the 29th both hope and ship were dashed. The day began encouragingly enough. The *Kaponga* had remained fast all night and, the next morning, the *Kaimai,* outside the harbour, and the *Komata* inside, pulled on their hawsers to try to bring the *Kaponga*'s bow upstream, so it could be towed into port. They moved the ship around to face upstream, but could not dislodge the midships section from the seabed. Wharfies started removing coal to lighten the ship but completed just 40 tonnes before the weather turned nasty. By 1800 hours big rollers were pushing the ship towards the fatal north tiphead.

The crew rigged up a breeches buoy to the north tiphead and by 0420 all were ashore, even though the ship's rolling had whipped the line up and down fiercely. Captain Gray was the last off. 'As his feet touched the ground,' the paper reported, 'the small group standing on the pile of rocks in the pouring rain, with acetylene flares showing up haggard faces, spontaneously cheered.' One sailor brought the ship's cat and her three small kittens in a pillowslip, tucked safely under his shirt. All found homes with rescuers.[5]

The morning's tide washed big seas over the ship, which now listed heavily. When the engine room flooded and the fore and aft bulkheads collapsed, the *Terawhiti* was sent back to Wellington. Salvage was impossible, and with the wreck up against the breakwater the channel was again clear, although temporarily restricted daylight navigation. In late October the wreck broke up. The funnel went overboard on 30 October and the stern disappeared, leaving just 10 metres of the bow above water. It was expected

The West Coast's maritime trade settled down to coal and timber exports through Greymouth and Westport. Yet, even though ships grew bigger and better, the rivers, particularly the Grey, continued to exact their toll.

to sink in the next fresh. Coal washing up onto the North Beach delighted householders.

The court of inquiry found that Greymouth harbourmaster Captain Frederick William Cox had overestimated the depth of water over the bar when sounding it that day. There had been only 5.5 metres. 'I can't fathom it at all', Cox punned unwittingly, but the explanation was simple: the soundings had been made over too narrow a section of the channel and had missed a patch where shoaling was building up.[6] The Marine Department ordered the harbour board to sound over a greater width of the entrance, to alter the leads, and to improve the signalling methods. The last big craft wrecked at Greymouth, the freighter *Abel Tasman,* did so under unusual circumstances. On 17 and 18 July 1936 it rained as it can only on the West Coast. The Grey River became swollen with muddy water and, by early in the evening of the 18th, threatened to top the wharf. This, and an unusually strong ebb tide, broke the *Abel*

Tasman's moorings. Captain W.D. Archibald got steam up and tried to beach at Blaketown Lagoon, but at the bight on the confusing currents, the ship drifted out through the entrance and was swept around to the rocks on the North Beach where it became a complete wreck. There were no casualties.

The one that got away — the *Mapourika*

The West Coast service was undoubtedly the most dangerous of the Union Company's passenger runs. Even the specialised shallow-draught steamers built in 1898 and 1905, the *Mapourika* and *Arahura*, had their work cut out to maintain a service to the shallow West Coast bar harbours, frequently barely scraping across the shallows, to the alarm of passengers. Often they lay 'bar-bound' off the river ports for days, unable to enter. Travelling was never entirely predictable.

The *Mapourika* (1203 tons) was one of two near sister ships built in 1898 for the company's West Coast passenger services; the other was the *Rotoiti* (1159 tons), which ran between Onehunga and New Plymouth. The *Mapourika* traded between Wellington, Nelson, Westport and Greymouth. It was a handsome single-funnel ship, with a straight stem and elliptical stern, 67 metres long, 10 metres in the beam and with a draught of just under 4.5 metres. Triple expansion engines gave a speed of a shade over 11 knots.

Because the *Mapourika* was almost twice the size of its predecessors it soon became a firm favourite with travellers such as Judge Ward, who used the steamers while making his district court circuit. Being so tall and fat, he had previously had to bunk down on smoking room floors previously. Now, Ward had a special berth fitted for him in the *Mapourika*.

At about 0100 hours on 1 October 1898 the *Mapourika* was lining up to enter Greymouth. Captain McLean, ex-*Mawhera* and *Penguin*, knew the West Coast harbours well. The first officer was Mr Liddell. Although the night was dark, the weather was fine and the sea over the bar only moderate.

At half-speed McLean came in line with the lights. A heavy sea was running, but not enough to trouble the *Mapourika;* there was plenty of water over the bar. Just as it was three ships' lengths from the southern tiphead, though, a huge 'blind roller' slammed into the vessel amidships. The timing was awful. With its helm and propeller out of the water, the *Mapourika* hung in the air for what seemed like a minute, then plunged, striking the seabed and becoming unmanageable. Mclean ordered full speed ahead but got no response. Then another big sea swept the ship towards the northern tiphead.

McLean ordered 'full astern', but nothing worked. Its propeller racing full astern, the *Mapourika* was carried onto the northern tiphead, slowly working up onto the northern beach. It struck so heavily that a 20-tonne concrete block was split in half.

Passengers reported just a brief moment

This photograph shows the *Mapourika* ashore near the tiphead, pounded by the waves. A brave (or foolish?) man is walking on the ropes that so tenuously connect the ship to shore.
Gift of Wellington Museums Trust, New Zealand Maritime Museum (2012.0.1630)

of panic as they made their way up on deck. Fortunately the ship's stern was so far up the beach that they scrambled ashore safely, carrying luggage and the mail. But the ship's future looked less promising. 'The fine new steamer *Mapourika,* which recently came out from Home to the order of the Union Steam Ship Company, had gone ashore after crossing the Grey Bar and was likely to become a total wreck,' an early wire report revealed. 'She now lies on the north side of the north groin, with her bows on the rocks ... She has holes in the bows.'[7]

So did the port's reputation. The *Mapourika* was a symbol of progress and Greymouth businessmen feared that the Union Company might never risk such a valuable ship there again. The *Grey River Argus,* sensitive to criticism of the port, called for an inquiry. It was held the following week and was surprisingly uncontroversial. Although the Greymouth harbourmaster, possibly overly anxious to protect the reputation of his port, 'considered the *Mapourika* unsuitable to that trade, because she drew so much water and had so much of her propeller blades out of the water', the magistrate found that the accident had been caused by the ship becoming unmanageable through striking the seabed after being hit by a blind roller.[8] Such rollers were not uncommon, especially in summer months; earlier one had practically turned the steamer *Janet Nicoll* around. The bar had been safe and the master perfectly justified in entering under those conditions.

Although many wrote off the *Mapourika,* Union Company branch manager W.A. Kennedy disagreed. He had rescued the *Mawhera* from the same place a few years earlier by dragging it across the beach and relaunching it in a more sheltered spot. This had taken months and cost a fortune, but had saved a valuable ship. Kennedy wanted to save this fine new ship and summoned J. Daniels, foreman of the company's workshops, to Greymouth to assist.

One thing was certain — the ship had to be moved fast. It was in a very dangerous position, with its submerged bow in deep water and its stern on the rocks. A devilish combination of fast-flowing river water and incoming heavy surf ruled out using tugs. Daniel ran out five hawsers to the stern and one to each side of the bow to swing the stern further from the breakwater. But he could do nothing about the bow that, in deeper water, was being pounded by waves. A heavy southwester sprang up on the 5th and caused the *Mapourika* to bump heavily. Fortunately, the engine room remained watertight and they could still raise steam.

Nothing went well at first. Daniel had requested pumps and additional men from Wellington, but bar conditions forced the ship carrying them to divert to Westport. In the meantime, Daniel had tried to move the ship using a steam winch but the ropes kept breaking.

With the aid of the new pumps, Kennedy started pumping out the forepart of the ship from the night of the 12th. Workers then shovelled out 150 tonnes of shingle from the fractured hull while others, assisted by divers, patched the holed port side. It was a race against time. On 17–18 October a heavy storm threw the ship further up the beach, twisting the stern and holing and denting more plates. But through a combination of steam and muscle power, Kennedy and Daniel hauled the *Mapourika* out of the water. By 20 October it was high and dry, and safe.

Daniel's men then removed the rest of the damaged cargo and repaired the hull damage. To get the *Mapourika* over to the other beach for launching, they started levelling a section of the beach and removing part of the breakwater wall. This took four months. Daniel then lifted the *Mapourika* onto wooden skids and manoeuvred it about 100 metres across the beach for relaunching.

On the big launch day of 10 March 1899, 6000 people watched the *Mapourika* slip faultlessly down the skids into the water. At the ceremony, James Mills — touchy about press criticism that his company had followed its usual policy of firing

By late October/early November the *Mapourika* had been winched up onto the beach, away from the waves. The next step was to build a wooden framework to enable jacks to lift the ship onto the wooden rails that would carry it across the breakwater for relaunching.

John Charlton Collection, Grey District Library

masters unlucky enough to lose their ships — replied that 'if the waters were difficult to navigate the company did not enforce that law [summary dismissal] but considered the circumstances. Where there was land on one side and 3,000 miles of open ocean on the other the matter was different, and in the interests of the shareholders and of the general public — whose lives and property were entrusted to them — the company could not act otherwise than they did.'[9]

This was not the *Mapourika*'s last accident. In August 1900 it ran ashore at Picton and was again the subject of a major salvage exercise. Other minor accidents occurred at regular intervals. In 1907 the ship had another close call at Greymouth and other incidents occurred elsewhere. Sold to the Anchor Company of Nelson in 1921 as the *Ngaio* and laid up in 1930, the *Mapourika* was eventually dismantled in the mid-1930s.

Most of the river ports have gone now, silted up as the consequence of agricultural clearances in their headwaters, competition from land transport and their inability to handle modern-size ships. Most went in the nineteenth century but a few lingered until comparatively recently, expiring in the 1950s (Whakatane) or the 1960s (Patea and Kaiapoi). After going through a dark period, first Westport and now Greymouth have revived thanks to coal (both ports) and cement (Westport). Whanganui's future is uncertain. Yet, as the continuing spate of fishing boat tragedies on the Grey bar shows, these ports remain perilous for the unwary or the unlucky.

By December the *Mapourika* had been jacked up by three metres and was being eased gradually across the beach. It was a massive task. The *Mapourika* was relaunched into the Grey River in March 1899.
Ref: PAColl-9097-6, Alexander Turnbull Library, Wellington, NZ

4

Exposed coasts — roadstead ports

For all their shortcomings, river ports at least provided shelter. But what about those settlements that lacked even token protection, towns such as New Plymouth, Napier, Timaru and Oamaru, born on exposed beaches open to the full force of the sea? They faced a stark choice: build a costly artificial port from scratch or fall behind luckier rivals.

It was never easy. Breakwaters, wharves and dredging had to compete for funds with other public utilities and services. There was never enough to go around, and sometimes politicians proved better at shifting paper than timber or concrete. While their leaders pontificated, pleaded and wheedled, people struggled as best they could. Sometimes they resorted to desperate expedients. Oamaruvians, who would watch the Pacific Ocean splinter their unprotected jetty in 1868, tried to put an improbable dock in the town's tiny lagoon. Further up the coast, Temuka proposed a dock in Milford Lagoon as an alternative to Timaru's troubled port. Napier delayed even longer, using the shallow lagoon at Port Ahuriri well into the twentieth-century. It took the 1931 earthquake to tip the scales in favour of the deepwater breakwater port begun in the 1880s and largely mothballed about 1909. New Plymouth's harbour history was barely less troubled.

Mariners always feared open roadsteads. New Zealand's coasts are lashed by some of the world's most changeable weather, which can catch out even masters keeping a 'weather eye' on the horizon. The early history of Oamaru and Timaru is littered with the wreckage of ships caught off a lee shore.

A dark and stormy night — Oamaru 3/4 February 1868

Our ancestors saw things very differently. After one of the most violent storms in New Zealand's recorded history swept along the South Island, flooding rivers, drowning farm workers and wrecking ships, the Otago provincial authorities ordered a day of fasting and public prayer. Today we would order an inquiry and the press would speculate about global warming. 'The late hour at which our notice was penned and the

OPPOSITE Although the brigantine *Emulous* looks intact enough in this photograph, storms soon battered it to pieces. Of 157 tons register, the *Emulous* had been built at Hartsport, Hampshire, eight years earlier.
North Otago Museum

Mariners always feared open roadsteads. New Zealand's coasts are lashed by some of the world's most changeable weather, which can catch out even masters keeping a 'weather eye' on the horizon.

For years this unusual flagstaff warned shipping of danger. It is the foremast from the brigantine *Robert and Betsy*, wrecked on the night of 5 April 1862 after parting its cable. After serving on the foreshore and then in a Tyne Street garden, it was re-erected in the 1990s near the yacht club slipway.

Gavin McLean Collection

circumstance that our reporter himself was well-nigh drowned, must be our excuse for so meagre details of a disaster the most serious which has ever visited this port,' the *Oamaru Times and Waitaki Reporter* apologised to readers on the morning of Tuesday 4 February 1868. It need not have bothered. With all the clean-up work facing them, few would have had time to pore over their papers.

The storm that swept the coast on 3 February 1868 buffeted anchorages from the Catlins in the south to Lyttelton in the north before striking inland. Howling east-southeast winds and pouring rain caused rivers from South Otago to North Canterbury to burst their banks, flooding farmland, sweeping away isolated settlers, demolishing bridges, damaging buildings and wrecking business premises. Many people, by no means all seafarers, died on 3–4 February.

It began innocently enough when the Oamaru beachmaster, Captain William Sewell, ran up the Blue Peter — the warning to head out to sea. Four ships were in the bay: the wool ships *Star of Tasmania* and *Water Nymph*, the ketch *Otago* and the schooner *Emu*. Laid down as a corvette for the French Navy, the *Water Nymph*, had been finished in 1855 as a merchant ship. The *Star of Tasmania*, meanwhile, was a fine, clipper-rigged ship. Built by Hall's Aberdeen yard 12 years earlier, for years it had been calling at Oamaru for wool for London. The *Star* had 22 aboard: the master, two officers, 10 ABs, one OS, the cook, carpenter, steward and boatswain, a woman passenger and her two children and, lastly, a boy named McLean. Before the day was out, the two Baker children and two seamen, David Petrie of Arbroath and Londoner William Brookes, would die in Oamaru's worst maritime disaster.

The smaller craft were better placed to escape. At 1330, half an hour after seeing the signal, the *Emu* spread canvas and worked out, hauling as close to the wind as possible. Half an hour later the *Otago* set its mainsail, staysail and jib and stood out.

The larger ships stayed put. It was later discovered that the *Star* could not make out the signals on the flagstaff and that it blocked the *Water Nymph*'s escape. In frustration, Sewell shouted to the *Nymph* (lying to leeward of the *Star*) to get away before it was too late, but the wind blew his words back in his face. At 1530 he ran up another signal to 'proceed to sea without delay'. Locals, alerted to the signs of impending disaster, gathered at the anchorage.

Then Sewell noticed the *Star of Tasmania* dragging its chains. The moorings offered little security. The *Star* had broken them on an earlier visit, as had other ships. The authorities relaid them, but few had much faith in them in bad weather. Not long before, the *Oamaru Times* had noted 'another instance of the absolute inefficiency of the present moorings in our Bay' when a southerly had caught the barque *A.W. Stevens* in the anchorage while Captain Brown was ashore. The barque's first officer cut the hawser springs and headed for sea, forcing Brown to chase after the *Stevens* in a boat. It took him three hours to reboard his ship. But his discomfort was slight compared to that of the women who had been visiting Captain Brown's wife. The winds and seas kept them from their homes for another five days.[1]

Midway through the afternoon the moorings snapped. The *Star of Tasmania* drifted about 400–500 metres before Captain William Culbert brought the ship up by dropping the starboard anchor. The forecastle party fished up the broken chain and then dropped the port anchor. It held, but only briefly. Both anchor cables parted with a jolt and the big ship drove shorewards. The crew tried to hoist sail but the huge rollers sweeping the *Star*'s decks tossed the deck cargo about, making it too dangerous to work the ship. They took to the rigging and braced for the final crash. At 1900 the *Star of Tasmania* fetched up broadside-on, opposite the Presbyterian Church, with its head to the southward. In the words of the reporter sheltering on the spray-lashed beach:

Captain Babat, master of the ill-fated *Water Nymph*.
White Wings

There was by now an awful sea on, the breakers being of a magnitude never before seen on this part of the coast. The vessel rolled to and fro upon the shingle, and being heavily laden, strained and creaked as the enormous masses of water struck her and knocked her about. In a very few minutes the copper on the starboard side was torn off her timbers, and the water pouring through her seams showed that the port bilge had been driven in. Wave after wave leaped clean over her, and the vessel finally fell over on her port side, her masts quickly afterwards falling into the sea.

Now the real drama began. Sensing that the *Star* would quickly break up, the passengers and crew crawled to the starboard bow. Some barely made it: Captain Culbert leaped off the *Star*'s poop just as it started to break up. The Baker children had already drowned in their berths. But the forecastle was a temporary refuge. The seas were crashing over the ship, flinging wool bales, iron tanks, spars and timbers at the terrified survivors.

Onlookers could offer only sympathy: 'to add to the miseries of the scene it was raining as we never saw it rain before, the water corning down in torrents and making it impossible to look towards the vessel', the paper reported. 'The spray, too, came over in great white sheets, and struck upon the faces of the hundreds who were gathered upon the beach in the hope of rendering assistance to the unfortunate crew, with a cutting violence which made it appear as though it were a shower of needles.'

Rescuers tried to get a line aboard, but each time the wind, now shrieking at gale force, threw it back, even though the *Star of Tasmania* was a mere 10–12 metres away.

After watching these failures, the mate, Stevens, leaped from the forecastle and struck for the shore. Onlookers saw him disappear beneath the water, then reappear when a receding wave revealed him on his hands and knees about halfway between ship and shore. Another wave was rearing up, bristling with timber and debris, when bystanders, judging the moment perfectly, rushed in and dragged him ashore, the sea sucking at their feet as they raced to safety. A huge cheer went up but Stevens screamed with dismay, 'The line! The line! I've lost the line!' He had gone over the side with a line but lost it in the surf.

Surfboats unloading cargo at Oamaru about 1864. Such scenes were common throughout early colonial New Zealand. Initially boatmen rowed between shore and ship, but before long they were hauled back and forth along a line running from beyond the surf to the head of the slipways on shore. This eliminated some of the sweat but still made boating a wet and dangerous job.
North Otago Museum

Two other crewmen who followed his example were dragged ashore more through good luck than anything. A fourth man misjudged the moment to jump, got caught in the undertow, was swept past the bow and carried out to sea. A fifth repeated his error and was dragged out past the bow clutching a wool bale. Shocked by these tragedies, onlookers shouted to the others to wait for a line.

But what line? Boatman George McKenzie had been trying to get one aboard. Again and again he waded into the boiling surf and hurled his line, only to see it fall short. But he kept trying until collapsing, completely exhausted.

As darkness fell, a dray load of firewood, oakum and turpentine was brought up to start an immense bonfire. Then police sergeant Bullen waded into the surf, a rope fastened around his waist. He struggled in vain, but came close with one throw that put the end of the rope on the forecastle. Unfortunately, in the darkness no one saw it before the waves swept it away. Then a wave knocked Bullen off his feet, almost washing him away.

Things looked hopeless, though no one wanted to admit defeat. Captain Steward of the Volunteers tried to fire a line to the ship, using an improvised rocket gun, a stout fishing line fired from a rifle. Unfortunately, the line separated from the ramrod. Meanwhile, others had dragged up a lifeboat long dismissed as useless. One look at the surf convinced them to leave the boat where it was.

George McKenzie, his breath recovered, resumed his efforts and was rewarded at 2240, when the rope hit the *Star*'s deck and a crewman grabbed it. Minutes later boatman Duncan Young clambered aboard to loud cheers from the beach. Stevens followed, returning with young master McLean, whose parents lived in the town, clinging to his shoulders. One by one, the other passengers and crew slid down the line to the beach.

Stevens returned for Mrs Baker, whom he slung over his shoulder. 'He then commenced descending the lifeline with his burden, but just as he reached the water an immense wave dashed over the vessel and hid them for a moment from view', the paper reported. 'As the wave retired Stevens was seen hanging to the rope with Mrs Baker still clinging to him. Again a wave submerged them, and it was feared that they must both be carried away, but on its subsidence, they were found to still be safe and in a few moments Stevens had got near enough to grasp the helping hands held out to him, and the lady and her preserver were brought safely ashore amid loud cheers.' The first words uttered by the brave woman were: 'Never mind me, save the poor, dear Captain.'

Last off was the hero of the night, Duncan Young, who, in Hollywood style, ran to the waist of the ship, jumped on part of the mast that was wedged underneath the keel and plunged ashore just as a wave retired, cheered by hundreds.

Other dramas had been taking place while this was going on. The *Water Nymph,* which had been moored closer to the shore, started dragging its anchor about half an hour after the *Star* struck. The *Star* had effectively blocked it until then, and now it was too late for the *Nymph* to escape. At about 1700 Captain Edwin Babot dropped his other anchor, but the seas were now too rough for any anchor cable to hold. Both cables snapped within minutes, so Babot put on all canvas to drive the ship as far up the beach as possible. It worked. The *Water Nymph* struck about 100 metres north of the *Star of Tasmania,* its bow pointing north, but far enough up the shingle to enable the crew to escape with all their effects after a line was put aboard.

The *Otago,* which we left clawing out to sea under close reefed canvas, also came to grief that night. The schooner seemed safe, but at 1730 hours its rudder carried away, leaving it unmanageable. Captain Campbell lowered the main sail and pointed his doomed ship at the shore, intending to run right up onto the beach. The *Otago* hit about 13 kilometres north of Oamaru, grounding bow-on as intended, but the waves pushed the ship over. The crew escaped with only the clothes they wore, while the *Otago* broke up, scattering coal and timber. In the morning all that could be found were a name board at Boundary Creek and a few spars and planks.

Dawn revealed a port in ruins. The jetty, only months old, was matchwood, its T-end missing and the approachway undermined. Most of the surfboats of the Oamaru Landing Service and Traill, Roxby & Co. were also ruined. The town was stunned. The inquiry cleared all three masters, but criticised Captain Culbert for improper management of his vessel and for beaching it end-on when he knew he could not save it.

Building a port

By 1868 the people of Oamaru had got used to shipwrecks. They knew that as a harbour, Oamaru left much to be desired. Cape Wanbrow, a bare, stubby little headland, provided some shelter from southerly winds but none from the predominant easterlies thrashing this exposed coast. When trade began in the 1850s, everything — pigs, pianos and people — had to be discharged into large, open surfboats. Capable of holding five or six tonnes, they would shoot through the breakers in a hair-raising surge of foam to the beach where the boat crews sledged them along the beach to a shed:

> A stout hawser rope through slots at the bow and stern of a boat called the tender which was moored several chains off-shore, was secured to stanchions above the beach. A surf boat returning from a vessel arrived alongside the hawser and the bow oars man would ship his oar and secure the hawser with a short boathook and hold the bow until the other oarsmen had swung the boat bow to seaward. Then the hawser would be shipped into slots at the bow and stern, and the crew would proceed to haul the boat hand over hand until the stern grounded on the beach. Then the boat would be firmly secured to the hawser so that the bow still remained afloat. By this time the shore crew would have waded out alongside and the cargo would be discharged. Having taken the return load the crew would grasp the hawser and a number of the shore crew would be ready to apply pressure to the stern of the boat as each wave lifted the bows; so that by united push and pull the boat was launched into deep water, the oarsman continuing to haul on the hawser until they arrived just behind the tender, when the hawser was cast off and the boat rowed to the waiting ship.[2]

People also took their chances. An early resident recalled being 'carried ashore through the surf by big strong boatmen, who waded out and took the passengers on their backs and carried them ashore'.[3] This was possible only in fine, calm weather. At the first sign of danger, or a shift in the wind, unloading halted, and ships scuttled out to sea. As trade grew, so did the number of ships playing Russian roulette off Oamaru's beach. The sensible ones usually unloaded during daylight and sheltered overnight at Moeraki: easterly gales made it too dangerous to lie off Oamaru in the dark. The Otago provincial government had licensed the boating service and put a derrick above the beach, but provincial harbourmaster William Thompson hated Oamaru and openly advised merchants to use Moeraki instead.

Port improvements were slow in coming. In November 1860 the provincial government, as bereft of sense as it was of funds, moored the hulk *Thomas and Henry* off the beach as a floating store. Fit for nothing, it merely added to the hazards of the anchorage and was towed away in February 1861.

In 1869 Captain William Sewell, the head of the landing service, organised the rocket brigade.

It took its equipment on a dray to the beach when required. Its main tool was a Boxer rocket gun, which shot a light line ahead of a thicker rope out to the ship whose crew rigged up a breeches buoy to provide a means of getting ashore.

The brigade had its busiest day on 27 August 1873 when two ships were dashed to pieces on the beach. The *Emile* was the first to go. Its fate had been sealed when the schooner *Jane Anderson,* fleeing the looming storm, hit the *Emile,* wrecking its jib boom and doing other damage that prevented it from escaping. While immense bonfires burning on the beach provided eerie light to work by, the brigade's men got a line to the *Emile* and rescued the crew, who were swiftly conveyed to dry lodgings. It had not gone smoothly. The first rocket line fouled the launching frame, disabling it, so subsequent shots had to be fired from an improvised triangle. The sixth, successful shot set fire to the *Emile*'s topsail before the line dropped into the crew's hands!

The master of the 231-ton brig *Scotsman* which, like the *Emile,* was carrying coal, had been hemmed in by the latter vessel. Anchored about two ships' lengths from the shore, in about six metres of water, the *Scotsman* bumped the seabed, then broke adrift. This time the rocket brigade got a line aboard very quickly, but for the people aboard, the trip ashore was terrifying. The waves were buffeting the *Scotsman* so badly that the rope, slackening and then tightening, sent one unfortunate woman yo-yoing high up into the air and then down into the water before she landed, dazed, drenched but alive. The *North Otago Times* praised the brigade but admitted that its management could be improved.[4]

In 1867 eight ships piled up on the beaches around town, four becoming total losses. The new schooner *Stately* went ashore when a heavy sea swept the anchorage on 14 March. Four ships escaped and the other casualty, the schooner *Vixen* was refloated. The *Stately* trusted to the moorings and was lost. The *Oamaru Times* did not mince its words:

By 1868 the people of Oamaru had got used to shipwrecks. They knew that as a harbour, Oamaru left much to be desired. Cape Wanbrow, a bare, stubby little headland, provided some shelter from southerly winds but none from the predominant easterlies thrashing this exposed coast.

The *Stately* was thought to have been in comparative safety, the sea having subdued considerably; and after 11 o'clock an attempt was made to clear out of port. This unfortunately was beyond his power, and about 12 o'clock Captain Short had the misfortune to find his fine craft going fast ashore, which she did close by the Landing Place, and, as ill-luck would have it, amongst some rocks which abound there ... Various opinions have been passed as to the course pursued by the master; many considering that he was perfectly justified in holding on to the moorings, they are being reckoned quite sufficient and secure for a vessel very much larger than the *Stately.* ... Be that as it may, we have the fact before us — the moorings have been found shamefully insecure.[5]

The wreck brought a mere £200.⁶

A few months later, on 31 July, another sextet was sent scattering by a southeast gale. Three escaped. Of the others, the cutter *Hope* and schooner *Midlothian* were beached but recovered. The principal casualty was the *Vistula*, which its master believed sufficiently well ballasted to remain at its moorings. He was wrong. At 1730 it snapped them and drifted ashore. The *Vistula* was righted and unloaded, but another storm wrecked the brigantine before it could be recovered. Just a fortnight later the schooner *Banshee* joined the casualty list. It was refloated, battered but safe; others would not be so lucky.

On 22 November 1867 a heavy gale raised a tremendous sea along the east coast of Otago. As usual, Oamaru got the worst of it. The old brig *Highlande*, heavily laden with New South Wales coal, parted its cables and hit the shore where it broke up. Then it was the turn of the local schooner *Caroline*. This little craft made it out to sea and looked safe until the gale split its sails off Kakanui. Unable to gain an offing, and flooding from wave damage, the *Caroline* was run ashore on Sunday 24 November to save lives. The ship grounded alongside the *Vistula*'s wreckage and quickly broke up. The crew escaped, but one man nearly died when he was dragged under the ship and was saved only by the quick intervention of onlookers. For once the spectators fared worse than the crew. A wave broke over the jetty, sweeping away two onlookers. One was rescued, but Daniel McLeod drowned.⁷

As we saw, 1868 was even more devastating. The tragedy forced a rethink. After political bloodletting and 'indignation meetings', Oamaruvians decided to build a dock in the

The 3rd of May 1874 was a bleak day at Oamaru with the schooner *Ocean Wave* (left) and brigantine *Emulous* (right) on the beach. Neither survived.
North Otago Museum

lagoon. The creek and lagoon were small and narrow but engineers believed that they could squeeze in a satisfactory wet dock. Protected at its entrance by twin sea walls (in much the same manner as training walls at river ports) and by a sea wall running off Cape Wanbrow, it would allow 300-ton ships to discharge cargo in the heart of the commercial district safely — in theory, anyway. Some work was done towards manufacturing concrete blocks for the dock, but after several changes of engineer, and some vigorous board debates, the dock trust decided to concentrate on the Cape seawall. The dock was quietly forgotten. Fortunately Oamaru's growing population (it was New Zealand's ninth largest centre in 1878) gave them the money they needed. Contractors started on the 564-metre-long breakwater, approximately in line with the outer sea wall of the dock scheme, and worked away on it until 1884, often in the face of appalling conditions and at considerable risk to the workers.

But it was worth it. Long before its completion, the breakwater was protecting shipping. Total wrecks fell away at the port. The *Premier* (wrecked twice in 1871, first on 31 July and then on 29 September) had been followed onto the beach in 1872 by the schooner *Margaret Campbell* and then, as we saw, by the brigs *Emile* and *Scotsman*. In 1874 the number of total wrecks was again three — the schooner *Ocean Wave* and brigantine *Emulous* on the night of 2/3 May and the schooner

The locally owned barque *Premier* ashore at Oamaru in 1871. The 296-ton vessel was insured for £1500, barely half its replacement cost.
North Otago Museum

United Brothers. The curious fate of the *Emulous* echoed that of the *Premier*. Stranded but refloated, it was prepared for the voyage to Port Chalmers for permanent repairs. Three times it ventured out to sea, only to be beaten back to Oamaru. On the fourth attempt, on 11 October, fate intervened for the last time. Forced back when just eight nautical miles from safety, the unlucky ship was pounded apart just three miles north of Oamaru.

By 1875 the breakwater had advanced far enough to offer reasonable shelter for most port visitors. With the opening of the first wharf (Macandrew) that year, fewer ships took their chances in the roadstead. As more were now powered by steam, their dependence on the whims of wind and tide diminished.

Shipping casualties declined dramatically. There was only one total wreck at Oamaru in 1875, the three-masted schooner *Elderslie* in May. No more were lost until June 1879, when the ketch *Franklin Belle* broke adrift while lying off Normanby Wharf and hit the beach. Four years later, Oamaru's last major wreck occurred when the schooner *Friendship* was lost while trying to enter port in bad weather. No more trading vessels were wrecked at Oamaru before the port closed to commercial shipping in 1974. Here, as at Timaru, New Plymouth and Napier, the breakwater port, though expensive, proved its worth.

Tragedy at Timaru

Timaru shared a coastline similar to Oamaru's and a list of troubles almost as long. Like their southern rivals, Timaruvians had dithered, debated and dallied while the wrecks piled up. Like Oamaru, too, Timaru was finally building an expensive artificial breakwater port, though a little behind its southern rival. In May 1882 some might have been forgiven for thinking that it was all a bit too late. Timaru had not been kind to shipping that year. On 14 January the big iron-hulled ship *City of Cashmere* parted its cable in seas so light that the crew initially did not notice. The steam launch *Lillie Durham* tried but failed to tow the ship to safety, and the *City of Cashmere* drifted ashore about six kilometres from the port. The rocket brigade rescued the crew, and the ship broke up the next day.

The next casualty was the 1047-ton barque *Duke of Sutherland*. The *Duke* had been stuck off Timaru since 25 March, loading wheat, but had been delayed by shortages of boats and labour. The ship's master complained to the harbour board about being left for so long in such a dangerous position, but his ship was still lying in the roadstead on the evening of 2 May 1882 when disaster struck.

Although the tide was low, the ship was drawing about 5.5 metres at an anchorage said to be 8.5 metres deep at low tide. Larger ships had used that spot safely. A heavy sea had been running all day, but as the ships anchored off the partly built breakwater were riding their anchors easily, all seemed well until shortly after 1700, when the *Duke* sent up blue lights and rockets. Harbourmaster Captain James Mills alerted the rocket brigade, but as the steamer *Waitaki* had passed the *Duke* shortly before without observing anything amiss, he was not too worried.

That complacency vanished an hour later when a boatload of men from the *Duke* pulled alongside the landing steps to report that at around 1700 an unusually big sea had dumped their ship's stern on the seabed. At first Captain Henry Rowlands thought they had been nudged by a floating object, and he was still looking for it a few minutes later when the *Duke* was struck again, this time heavily. Now there was no doubt. Rowlands flew distress signals. The *Ben Venue* sent across a boat, whose occupants were told that the impact had carried away the rudder post. The *Duke* was settling by the stern and by 2000, when most crew were ordered into the boats, the water was up to the 'tween decks.

Mills rowed out to see what could be done. Harbour board diver William Collis opened the lazarette hatch and discovered 'a considerable quantity of water in the ship', although he could not find the source of the leak. While Collis was exploring, Rowlands and Mills got the pumps working again. For several hours they kept up with the inrush of water. Mills did not want the *Duke of Sutherland* to sink in the middle of the port's entrance, so at 0400 he ran a line to the *Ben Venue* to have the *Duke* hauled north, though this soon began to slip, leaving the *Duke* drifting towards Caroline Bay — thankfully clear of shipping. It went ashore almost half a kilometre from its original anchorage. In just a few hours the sea wrecked the ship: the deck collapsed, the ship toppled onto its starboard side, and its main topmast and most of its yards collapsed. Rowlands salvaged spars, boats and other valuable items.

The *Timaru Herald* commented that: 'The mishap to the *Duke of Sutherland* is a most unfortunate one, coming so soon as it does after the wreck of the *City of Cashmere*. We are glad to learn that our Harbor [sic] Master has, from first to last, objected to large vessels being brought so close in, and that only a few days ago, although urgently pressed to do so, he refused to bring the ship *City of Perth* nearer the Landing Service than she at present lies. The hardship in the present case, so far as the vessel and the port is concerned, is that she was like the *City of Cashmere,* all but ready for sea, and but for the want of working appliances on the part of the Harbor [sic] Board, would ere this have been on her way to England.'

Twelve days later it was the turn of the *City of Perth* and the *Ben Venue*. Early on the morning of

Timaru shared a similar coastline and problems with Oamaru. This photograph shows the wrecks of the schooner *Duke of Edinburgh* (foreground) and the brig *Fairy Queen* (214 tons, 1863), both of which were swept ashore in heavy seas and strong winds on 27 August 1873, along with the ketch *Wanderer*. The *Duke* was badly damaged, but was ultimately refloated two weeks later. The others were not so lucky.
Ref: 1/2-019960-G, Alexander Turnbull Library, Wellington, NZ

Timaru initially depended on a landing service from its exposed beach. The George Street landing of the Timaru Landing and Shipping Company was established in 1868 by local merchants dissatisfied with the government service. This photograph, taken about 1873, shows the boats and a beached steamer. The edge of the bluestone Landing Service building can be seen at the extreme right.
Ref: 1/2-005348-F, Alexander Turnbull Library, Wellington, NZ

14 May conditions at the port started to deteriorate, showing the old pattern that locals knew spelled trouble — big seas running with huge rollers showing far out to the horizon. To make it worse, the day remained calm and sunny. The 999-ton *Ben Venue* caught lying stern-on to the sea, was soon shipping big rollers aft which caused the cargo to shift and the ship to list to starboard. It was in serious trouble. At 0900 a third anchor was dropped after the second parted. Four hours later both cables broke and Captain MacGowan ordered the crew to abandon ship. An hour later the *Ben Venue,* its bow still pointed shorewards, struck under the cliffs at Caroline Bay, where it turned broadside-on, its shattered decks facing seawards.

The crew rowed to the *City of Perth,* which proved a very temporary refuge when its cables also parted. Both crews took to the boats, leaving the *City of Perth* to ground alongside the *Ben Venue* on an even keel. For the thousands who had gathered on the cliffs to watch the breakers sending spray flying over the two big ships' rigging, it was a sight they would never forget.

The real tragedies had been acted out in open boats amongst the breakers. Mills disapproved of the decision to abandon the *City of Perth* and, seeing an improvement in the weather, set out to save it. Accompanied in a separate gig by the master of the *City of Perth,* Captain C. Macdonald, Mills headed out in a whaler, intending to halt the ship's drift. Unfortunately, their timing was out. Almost as soon as they scrambled onto the ship's deck, the last cable parted. Admitting that it was hopeless, Mills ordered some men into one of the

City of Perth's lifeboats and then directed the three small craft back to shore.

Just short of the lee of the breakwater, a heavy sea swamped the lifeboat and threw most of the men into the water. Mills steered towards them but was himself swamped. The third boat kept clear of trouble a little longer but eventually it, too, capsized. Now there were almost 40 men struggling in the mountainous seas. The lifeboat *Alexandra* was launched and made for the upturned boats at full speed. Its dramatic mercy dash, which involved three capsizes and rightings, was the stuff of Victorian melodrama. Eventually, the *Alexandra* drew alongside the men struggling in the water and began plucking them from the sea. But for some it was too late. The spills had cost the lives of two *City of Perth* crewmen and five Timaru watermen. In addition, two seamen suffered serious injury, and Mills died of injuries later that day.

That was the last such wrecking at Timaru. The breakwater was completed a few years later and the casualty rate fell away to almost nothing. The nearest thing to a serious accident came on 12 November 1964 when the freighter *Treneglos* (9976 tons, 1963) ran aground 1¼ miles south of Timaru. It had just left the South Canterbury port for Dunedin when it struck rocks just north of where the *Elginshire* (4579 tons) had gone ashore in March 1892. For the next five days the *Treneglos* lay pinned to the rocks until pulled clear. The damage took months to repair.

OPPOSITE TOP Like Oamaru, Timaru built its future with an expensive breakwater. This late 1878 photograph shows preliminary work getting underway on the second breakwater, built to replace an earlier (1870) groyne destroyed in heavy seas. In the meantime, ships lie off the exposed beach.
Ref: 1/2-005345-F, Alexander Turnbull Library, Wellington, NZ

BELOW One of the most dramatic pictures ever taken of a shipwreck in New Zealand. Breakers batter the *City of Perth* and *Ben Venue* in Caroline Bay, Timaru, on 14 May 1882, as onlookers gather atop the cliffs to watch the drama.
Gavin McLean Collection

TIMARU'S LIFEBOAT

In 1859 the Canterbury provincial council bought a 7.6-metre lifeboat for Timaru, where it deteriorated on the beach until a boatshed was built for it in 1862. By then the damage had been done, so the government asked John Marshman, its British agent, to recommend a replacement. Unfortunately, he rejected the Royal National Lifeboat Institution's standard designs in favour a 10.4-metre lighter, narrow (1.9-metres) craft from a boatyard at Limehouse. It had buoyancy chambers fore and aft, side air cases and a heavy keel; six brass self-relieving valves and tubes under the deck removed any water that came aboard.[8]

The boat and a horse-drawn boat carriage were shipped aboard the *Huntress*. The *Alexandra*, as the boat was named on arrival in 1863, was manned by volunteers from the Timaru Landing Service. It rescued two men from the *Prince Consort* in 1866, but its instability led to the death of a man from the *Susan Jane* in May 1869. After that the craft was not used again until 1882, when it was launched after open boats inspecting the *City of Perth* capsized, as did the *Alexandra*.

Three years later the harbour board stopped paying the crew's allowance and the *Alexandra* was stored in the open. In 1932 it was donated to the borough council, which displayed it under a shelter at Caroline Bay until 1996 when the new Timaru Maritime and Transport Trust took it over, putting the restored boat and a replica carriage on display in the restored Landing Service building three years later.

The restored *Alexandra* in front of the Landing Service building.
Phillip Brownie

PART TWO
The Sail and Steam Era

5

Hot times in cold climes — icebergs and castaways

'Now we know what happened to the *Dunedin*', Captain Cowan of the ship *Wellington* said in 1893. In the early morning darkness his ship, carrying a cargo of frozen meat from Picton to London, crashed into an iceberg 'with more mass than Rangitoto' in the frozen Southern Ocean. The ship survived, but not all crewmen did. One young seaman, asleep in the forecastle, probably died without knowing he was being crushed flat by the ship's crumpling iron hull plates. A less fortunate shipmate stayed conscious for four hours, horribly mutilated and pinned beneath the *Wellington*'s mangled bows before dying. 'For heaven's sake, Andy, get a gun and put me out of my agony,' he pleaded in vain.[1]

The *Dunedin* that Captain Cowan referred to was built by Robert Duncan at Glasgow in 1874. After several years as an immigrant ship, the *Dunedin* was selected to take a trial first shipment of New Zealand frozen meat to Britain. Much rested on the success of this voyage. If it succeeded, farmers could export meat as well as wool from their sheep, along with other meat and dairy products. Fitted with a Bell-Coleman cold air plant, the *Dunedin* loaded the historic cargo at Port Chalmers. There was a hiccup — when some machinery failed, one load had to be dumped on the local market before loading was completed — but the problem was fixed, and the *Dunedin* loaded more meat from the Totara Estate (outside Oamaru). The ship sailed from Port Chalmers on 15 February 1882 and delivered the cargo to London 98 days later. The success of this experiment was the making of the modern New Zealand economy.

After the Oamaru harbour works were completed, the *Dunedin* called there regularly, loading frozen meat, rabbit skins and other general cargo. In 1890 the ship was still in fine order. It arrived at Oamaru in January 1890 and, after a series of postponements, left in the early hours of 19 March 1890, towed out by the collier *Wareatea* to catch the wind for another voyage to London direct. At Oamaru the *Dunedin* had loaded 8848 mutton carcases, 3544 lamb carcases,

OPPOSITE Crew survey the damage of the *Wellington*'s bows, mangled in a collision with an iceberg in 1893.
Frederick G. Layton Collection, Ref: PAColl-6407-71, Alexander Turnbull Library, Wellington, NZ
PAGES 72–73 On the last day of October 1901 the Italian barque *Antiocco Accame* struck Danger Reef off the Shag River in North Otago. The 1106-ton ship had been built at Pertusola, Italy, ten years earlier.
Ref: 1/1-001042, Alexander Turnbull Library, Wellington, NZ

> **There were many dangers: fierce storms, the danger of collision with other ships in fog and, of course, icebergs. Hitting an iceberg could be fatal.**

1122 bales of wool, 205 casks of tallow and 1768 sacks of wheat, in addition to a small quantity of cargo taken aboard at Dunedin earlier. 'We wish Captain Roberts a safe and rapid trip Home,' the *North Otago Times* said on sailing day.[2]

The *Dunedin*'s voyage was neither safe nor rapid, for it never reached London. The loss puzzled those who had watched the ship loading at the North Otago port. The harbourmaster, Captain William Sewell, who knew the ship and its master well, considered the *Dunedin* fully seaworthy. In fact, he later recalled that Captain Roberts had told him the day before sailing that the ship had never been better, carrying 80 tonnes more stiffening than on previous voyages.[3] But something had destroyed the *Dunedin* and its 34 crew — what? Contact was made with the vessel once before reaching Cape Horn, after which it disappeared from sight. Did the *Dunedin* sink in a storm or hit one of the icebergs reported by other ships? Most mariners blamed the ice, which was thick that autumn.

Passengers and crew rightly feared icebergs. In the early 1850s enterprising skippers on the Australasian run had started taking their ships down into the previously empty southern waters to take advantage of favourable winds. In 1847 John Thomas Towson, a scientific examiner of masters and mates at Liverpool, published a small but influential book in which he argued that if the shortest distance between any two points on a sphere is a curve, it followed that ships should follow a circular route on long distance voyages. Three years later Captain Godfrey, in the *Constance* tested the theory by swinging further south — taking the so-called Great Circle Route — where favourable winds gave him a record passage to Adelaide. Others hesitated about using these cold, dangerous waters until the 1851 Australian gold rush produced a demand for fast voyages by gold-hungry prospectors. Almost overnight the icy-cold southern route became the normal one for anyone sailing to Australasia.

Down in these turbulent southern latitudes, mariners ran for up to 2000 miles when the ice permitted. There were many dangers: fierce storms, the danger of collision with other ships in fog and, of course, icebergs. Hitting an iceberg could be fatal for a wooden sailing ship. There were enough of them to avoid some seasons. Many migrant ships reported terrifying encounters with 'continent-sized' icebergs and fields that seemed to stretch to the horizon in all directions. When fog-bound in ice-infested waters, masters monitored the water temperature (many believed that low temperatures betrayed the presence of 'bergs) and doubled the lookouts.[4] Even so, they were a constant threat in such latitudes.

Some years were worse than others — the period between 1891 and 1897 produced a 'remarkable outburst of icebergs' in the waters south of New Zealand. In October 1892 the steamers *Coptic* and *Aorangi* reported many icebergs within a day's steaming of Auckland. A few days later, a ship that had left Lyttelton was 'completely surrounded by icebergs, nothing but ice three hundred feet high could be seen from aloft'. The ship, the *Star of India,* shut down its engines for a while. At Chatham Island that month, several icebergs drifted past the island, and some stranded at Pitt Island.

LEFT, TOP TO BOTTOM 'The ship *Dallam Tower* has been visited by thousands. She presents a picture of utter desolation, and it will cost about £6000 to place her in a state of good repair', the *Otago Witness* reported in September 1873. Three weeks earlier it had copied a strange report from a Melbourne paper: 'A ship was beating about Cape Otway in distress all day yesterday proves to be the *Dallam Tower*, from London, Captain Davies, for Otago. She is now being towed up the Bay, under jury masts, having lost all masts, sails, and even signal flags. She left London on 11 May. A steam tug left her off Seaford next day ... The *Dallam Tower* encountered a fearful hurricane when 3000 miles from Australia. She lost her masts by the board, and her decks swept of boats, galley — everything in fact. None of her cargo has suffered in the slightest degree. No casualties. Six passengers transhipped in a Sydney bound vessel that came up with the *Dallam Tower* soon after the accident.'

Mariners rightly feared dismasting at sea. These cards prepared by Port Chalmers photographer D.A. de Maus depict the story of one of the most famous examples of dismasting, the *Dallam Tower* (1499 tons, 1866). In mid-July 1873 the seven-year-old, 1499-ton ship was en route from London to Port Chalmers when a hurricane dismasted it, stove in its hatches and threw it on to its beam-ends. All navigational instruments were washed overboard, along with most deck fittings and the crew's gear. But, under the command of the enterprising first mate, George McDonald, the crew manned the pumps and rode out the storm, fashioning three crude jury masts from spare yards and wreckage. For many desperate hours it was literally all hands to the pumps for the passengers and crew while seamen cut away the rigging and wreckage. Fortunately, the winds, previously so destructive, now freshened behind the *Dallam Tower*, which limped into Melbourne on 19 August after making an epic 2000-mile voyage under jury rig. Officers and saloon passengers lost everything but the clothes they stood in. After lengthy repairs at Melbourne, the *Dallam Tower* finally reached Port Chalmers on 4 March 1874, more than six months behind schedule. The 13 passengers bound for Otago had crossed the Tasman earlier aboard the steamer *Albion*.

Ref (from top to bottom): PAColl-1/2-012591-G; PAColl-1/2-012592-G; PAColl-1/2-012593-G, Alexander Turnbull Library, Wellington, NZ

Typical was the experience of the renowned *Margaret Galbraith*. Home-bound in 1893, it had entered warmer waters and the crew had relaxed its guard, despite the squally, misty weather. Suddenly, the lookout spotted the weak sunlight glinting off a huge iceberg in front of the ship. The captain promptly swung away, but a seaman recalled that 'as we swung up into the wind you could have jumped from our quarter on to the ice that had so nearly been our doom'.[5]

When the mist cleared, the ship found itself in the midst of a huge field of icebergs of all shapes and sizes. How it had avoided hitting them no one knew. For the next four nerve-wracking days the *Margaret Galbraith* threaded through a field that included ice pinnacles nearly 100 metres tall. The biggest was thought to be 75 kilometres long and 300–350 metres high. Other ships encountering these dangerous floating islands that season — the *Turakina, Loch Torridon, Cutty Sark, Brier Holme* and *Charles Racine* agreed with these estimates.

Other New Zealand traders had similar adventures. The *Electra,* caught in an ice-field in 1869, could see nothing but mountainous icebergs in all directions. In early 1895 the ironically named *Himalaya* had to dodge them for almost 2000 miles. The *Matoaka,* which avoided them successfully in 1867 on a run from London to Lyttelton, is believed to have gone down after striking one two years later on a return voyage from Lyttelton.

The figurehead of the *Derry Castle*, wrecked on Enderby Island in March 1887. The *Southland Times* described the *Derry Castle* saga as 'a thrilling story of the sea', with the ship running aground on Enderby Island, part of the desolate Auckland Islands archipelago, nine days into a journey. For six months those not dashed to death on the rocks survived on shellfish and a few meagre supplies that had washed ashore, while sheltering in the remnants of a tiny hut earlier erected by the castaways of another wrecked ship, the *Invercauld*. The survivors were rescued by the sealer *Awarua*, which as luck would have it had called to pick up a dinghy that had been left on a previous trip.
Ref: 1/2-038208-G, Alexander Turnbull Library, Wellington, NZ

Castaways in the Southern Seas

Not all southern sea disasters involved hitting icebergs. There are rocks down there, too. The fierce storms that sweep the southern seas probably overwhelmed some ships that were reported missing. Others we know were wrecked on the sub-Antarctic islands that cover the south and south-eastern approaches to New Zealand — far from today's shipping routes, but lethal barriers in the days when sailing ships chased favourable winds along the Great Circle Route.

The shipwreck history of the sub-Antarctic islands is as murky as the weather that shrouds them. All that can be said for certain is that there were around a dozen *known* shipwrecks in the New Zealand sub-Antarctic islands between 1833 when wreckage from one or more ships was found, and 1908 when the *President Felix Faure* hit the Antipodes Islands. Although the loss of life was not, with the exception of the *General Grant*, especially high, the islands' remote location and the privations endured by survivors fascinated the public and made these wrecks special.

In a wreck on the mainland, rescue took hours or a day or two at the most. In the Aucklands or the Antipodes, however, it was merely the first scene in a very long story. The second part, the story of survival, could last for months or years. Typically, it involved struggling for survival, battling starvation and exposure, avoiding personality clashes and exercising human inventiveness to the full. For the desperate wretches living life on the edge, survival came down to outwitting dangerous sea lions or protecting their last precious matches. No wonder readers lapped up their stories.

François Raynal's story of the wreck of the *Grafton* and the five castaways' survival on Auckland Island helped make such stories a recognisable part of maritime literature. Raynal's account, first published in French in 1870, and in English in 1874, inspired Jules Verne's *Mysterious*

The remains of the *Grafton* were still evident in 1907 when the government steamer *Hinemoa* made a routine call to check and restock depots.
Ref: PAColl-3060-047, Alexander Turnbull Library, Wellington, NZ

Island. On a publication (two books) to survivor (five men) ratio, the *Grafton*'s contribution to castaway literature is exceptional. Captain Thomas Musgrave also kept a journal, which was also published, initially in London in 1866 and lastly in Wellington in 1943, edited by A.H. and A.W. Reed, as *Castaway on the Aucklands*. The Reverend H. Escott-Inman's book *The Castaways of Disappointment Island* (1911) is another classic of sea literature that has been reprinted often. Books have also been written about the *Grafton,* the *General Grant* and the *Invercauld* and, of course, the quest for the *General Grant* and its bullion still tempt treasure seekers (and writers).

'The land was soon lost sight of and I went to bed. But as I had not fallen asleep I heard the man on the lookout give the cry of "Land on our starboard bow"', survivor James Teer was reported as saying in the *Otago Witness* of 25 January 1868. 'While … was below the captain had hauled her on her course again. The land had the appearance of a fog bank, and it was on out lee beam, about three or four miles distance. The wind was fast falling away, and in a few minutes, it was dead. The captain did all in his power, with every flaw of wind from the flapping sails, but his attempts were useless. The yards were hauled in every possible direction that might enable the getting his ship off the shore, but all to no purpose, as the heavy S.W. swell was constantly setting her nearer and nearer the fatal rocks.

About 12 or 1 a.m., the ship was close to shore, and the current seemed to be setting her northward along the coast, until a rock stopped her progress. She touched it with her jib-boom and carried it away. She then shot astern to another point, which she struck with her spanker boom and rudder, injuring severely the man at the wheel. It was just half-past one a.m. on the 14th. The two points struck formed the entrance to a cove, and her side was rubbing against the perpendicular rocks. Owing to the darkness, we saw nothing save the dark mass above and around us. We could see overhanging rocks, and no place where a bird could rest upon them. Soundings were taken, and I think it was twenty-five fathoms under the stern, and all the while she kept working into the cave. The boats were then thought of, but the captain finding her lying so easy, and pieces of spars and rocks coming continually down, made it dangerous to attempt getting them out until daylight.'

This famous engraving from the *Illustrated London News* shows a contemporary reconstruction of the *General Grant* being swept into the sea cave.

Gavin McLean Collection

A curious feature of sub-Antarctic wrecks was that they occurred in clusters, separated by a decade or two. Two vessels were wrecked there in 1864 and one in 1866. These were the wrecks that drew the authorities' attention to the privations endured by wreck victims in such a bleak, isolated environment. Then, after a comparatively quiet period, the early 1890s saw the loss of two ships. Finally, in the early 1900s, after nearly a decade, the *Anjou* and the *Dundonald* were lost in 1905 and 1907 respectively.

The *Grafton* wreck was in some ways the least unfortunate. The small Sydney brigantine commanded by a young American master, Captain Thomas Musgrave, stopped at the Auckland Islands after prospecting unsuccessfully for tin at Campbell Island. The stop-off was to catch seals to defray the expedition's fixed costs. It proved a mistake, for on 3 January 1864, while lying in Carnley Harbour, a storm drove the *Grafton* onto rocks. The five crewmen escaped and were lucky that the wreck, well inside the harbour, stayed relatively intact, enabling them to salvage a boat, cooking equipment and the timber to build a hut.

Like later castaways, the men had to live off the island's limited resources, principally shellfish, fish, birds, seals and sea lions. 'The seals not only provided us with food and clothing but assisted me to make the journal from which this story of our adventures has been written up', Musgrave recorded in his journal. 'The small quantity of ink I procured from the schooner came to an end, and I found a substitute in seal blood.'[6]

Musgrave and François Raynal, the Frenchman who had conducted the mining survey, were the leaders of this small band. Raynal was a born survivor, having already survived cholera, temporary blindness and burial in a mine shaft collapse while gold-mining in Australia.[7] For over a year the men lived off seabirds and marine mammals, the only break to the monotony of life at 'Epigwaitt' ('dwelling by the water'), their campsite, being regular treks to the top of the hill to look for passing ships.

How long could they wait? When should they leave the illusory safety of life on the island for the obvious risk of sailing to New Zealand? It was not easy, but eventually, after realising that Sydney was not going to send a rescue expedition, and sensing that the island's meagre resources were not reliable (birds were migratory, and seals grew more wary over time), Raynal persuaded them to seek help. So, they built a forge and, led by Musgrave and the resourceful Raynal, started boatbuilding, lengthening, heightening and decking over their precious ship's boat.

That boat was too small to take everyone, so Henry Forges and George Harris remained on the island while their companions sought help. With mixed feelings, Musgrave, Raynal and Alexander 'Alick' Maclaren sailed for New Zealand. They lacked a chart or compass and the boat had to be bailed all the way, but Musgrave was a skilled mariner, and five stormy days later they reached Port Adventure, Stewart Island, where Captain Cross of the schooner *Flying Scud,* took them in. When the provincial and central governments refused to send a ship to rescue Forges and Harris, Southlanders raised money and sent the *Flying Scud* to the Aucklands, where they picked up their old shipmates and also discovered the body of the second mate of the *Invercauld,* a large ship wrecked on the western cliffs of the Aucklands, a few months after the *Grafton.* The *Invercauld*'s experience had been appalling. Six crew died in the wreck and 16 of the 19 who made it ashore died before the survivors were rescued by chance on 20 May 1865 when a Spanish brig stopped to repair a leak. Some men had been driven by desperation to eat the body of a dead shipmate.[8]

Only a year after the *Grafton* and *Invercauld* survivors reached New Zealand, the most notorious sub-Antarctic wreck occurred. The 1183-ton fully rigged ship *General Grant* was on its way from Melbourne to London with a general cargo and a large quantity of gold. Aboard were 83 people — 22 crew and 61 passengers. The voyage had started well, so there was shock and consternation at 2300 hours on the night of 13 May when the ship crashed into towering cliffs, carrying away its jib boom. Passengers and crew stumbled up on deck, dazed and confused, as the ship's terror ride continued. After drifting astern for about a kilometre, the *General Grant* struck again, losing its spanker boom and rudder.

That was just the prelude to the *General Grant*'s terrifying and protracted death ride. The ship's fate was sealed when the sea swept it bow-first into a huge cave, estimated by survivors to have been at least 80 metres deep. The ship's fore topmast hit the cave's roof and came tumbling down, as did the tops of the other masts, and the bowsprit was smashed off against the wall of the cave. The *General Grant* was now a battered, trapped hulk and as it worked its way further into the cavern, contact with its roof brought down more spars and rigging, forcing passengers and crew to shelter aft.

At daybreak the crew began getting people off,

but they got away just eight people in two small boats before the top stump of the main topmast — again striking the cave roof — collapsed. As the mast tops shortened, the sea carried the ship further and further into the cave, snapping off more mast timber as the ship bumped against the roof on each wave. Something had to give, and a series of bumps drove the footing of the main mast through the ship's hull. Water poured in and the *General Grant* sank rapidly, taking Captain William Loughlin and others with it. The longboat sank and many passengers, flung into the water, or who leapt from the sinking ship, drowned in the cave, whose walls, almost perpendicular in places, offered no place to land.

Only 15 survived, the people in the two boats launched before the masts crashed down and a very few strong swimmers who had escaped the wreck and were picked up. Just two — Fred Caughey and James Teer — were passengers. Bartholomew Brown, the first officer, was the only officer to survive.

It took time to find a safe landing. Fortunately, one man had guarded a match, and keeping alive the flame they lit with it became the castaways' priority during their long imprisonment on the island. After braving terrible conditions, they came to the same conclusion as the *Grafton*'s men — someone had to go for help. Who knew how long the bird and seal supply might last? On 22 January 1867, therefore, First Officer Brown and three men left in the ship's pinnace, which they had decked over with sealskins. Unfortunately, they were not as lucky as the *Grafton*'s men's. The boat, a shade over seven metres long, was never seen again and is presumed to have foundered at sea.

Ignorant of the fate of their companions, the 11 men left on the island endured considerable privations, and one of them, David McLellan, got sick and died. Shortly before, they had sighted a ship, but it failed to see the men's beacon and sailed past the island without stopping. Rescue finally came on 21 November when the whaling brig *Amherst*, engaged on a sealing expedition, anchored off the island and picked up the men.

The *Grafton*, *Invercauld* and *General Grant* incidents forced the government to build a chain of depots on the islands stocked with food, clothing, tools and small boats. Unfortunately, there was no depot on the island for one of the last ships to leave its bones on the Aucklands, the *Dundonald*. The huge four-masted barque, carrying Australian wheat, hit the aptly named Disappointment Island stern-first in the early hours of 7 March 1907 through a navigation error. Sixteen of the 28 crew survived, scrambling ashore as the masts grated against the cliffs.

Even safely ashore, the men's position was

The *General Grant*'s gold still fascinates us. Bill Day, chief executive of Seaworks, was photographed in 1996 with coins recovered from the sea floor near the Auckland Islands. In 2007, believing that up to $25 million worth of bullion might remain aboard, Day bought a $25,000 magnetometer and was planning to head south again.
Ref: EP/1996/1329/12, Alexander Turnbull Library, Wellington, NZ

precarious. Disappointment Island is tiny — just over three kilometres long and under two wide — and is a waterlogged, peaty hole of a place. To make matters worse, the fact that the ship sank quickly in rough, deep water meant that they salvaged little more than the sails. These provided welcome clothing and shelter, and one sailor had a small supply of matches, enabling them to cook the mollymawks and seals that formed their monotonous diet. The *Dundonald* men had many frustrations. The mate, Jabez Peters, died of his injuries after 12 days; a ship passed without seeing their bonfire; and every day they were taunted by the sight of the much larger Auckland Island, on which they knew supply depots existed. But as they had no seaworthy boat and no wood to build a raft, the big island was out of reach.

An inventive, industrious lot, they built coracle-like craft, using needles shaped from bird bones to sew sail canvas around wooden frames. A first expedition failed to scale the cliffs opposite and returned to Disappointment Island, but the men constructed other boats, and two months later, weak and starving, four men finally landed on Auckland Island. After gathering their strength, they followed a fingerpost that pointed the way to the depot at Erebus Cove. There they found food, equipment, three-piece suits and, most importantly, a whaleboat, which they used to row back to pick up their companions. The depot supplies included tinned meat and biscuits, and guns and ammunition, which were used to hunt cattle on the nearby Enderby and Rose Islands.

Greatly encouraged, and with their strength returning thanks to their improved diet, the men built a jetty and a flagstaff to signal the government steamer which they knew would eventually return to inspect and restock the depot. On 16 November they were rewarded by the sight of the *Hinemoa*. Captain J.P. Bollons gave them tobacco and tea and inspected the Campbell and Antipodes Islands before returning to pick them up and take them back to New Zealand, the second-to-last ship crew wrecked in unforgiving seas.

Sailors painting the castaway depot at Station Point, North East Island, Snares Island, c.1905. The depots were stocked and visited routinely until 1927. The small notice on the left warned against taking seals in the off-season.
Ref: PAColl-4660-01, Alexander Turnbull Library, Wellington, NZ

The *Dundonald* men were fortunate in that when they finally reached the Enderby depot, they found it intact and fully stocked. Not all were. At first it was assumed that distance would protect the supply depots from the depredations of casual vandals, and authorities did their best to discourage seal poachers, passing sailors and others from pillaging their contents. After poachers vandalised depots in the early 1880s the Marine Department sent the schooner *Kekeno* to inspect them between 1882 and 1886.

The Marine Department also tried to make their contents unattractive to all but the truly desperate, producing, for example, thick woollen suits of a distinctive pattern 'designed to ensure that the suits would always be recognisable and, therefore, would not be stolen by passing sealers', which can still be seen at the Museum of New Zealand in Wellington.[9] On one occasion an official on a depot relief expedition added a

PRINCIPAL SUB-ANTARCTIC WRECKS

Antipodes Island

Spirit of the Dawn	4 September 1893	After 87 days 11 of the 16 crew were rescued.
President Felix Faure	13 March 1908	22 crew survived and were rescued after two months.

Auckland Islands

Unknown ship	1833	Wreckage discovered by sealers.
Grafton	3 January 1864	All five crew survived. After 19 months on the island, three sailed to Stewart Island in a boat they had made and summoned help for their companions.
Invercauld	10 May 1864	Nineteen of the 25 crew reached shore. Three survivors were rescued a year later.
General Grant	14 May 1866	Of 83 on board, 15 reached shore. Four died trying to sail to New Zealand. Ten rescued after 18 months.
Compadre	19 March 1891	Following a fire, 16 of the 17 men were rescued on 30 June 1891.
Anjou	5 February 1905	All 22 survived and were rescued three months later.
Dundonald	7 March 1907	Sixteen of 28 were rescued 8 months later.

Enderby Island

Derry Castle	20 March 1887	Fifteen of 23 crew drowned. Survivors wre rescued 92 says later.

Macquarie Island

Kakanui	January 1891	Lost with 19 passengers and crew after leaving Macquarie Island.

message to a case of provisions to deter the light-fingered: 'The curse of the widow and fatherless light upon the man that breaks open this box, whilst he has a ship at his back'.[10] Unfortunately, the poachers were either illiterate or indifferent to supernatural powers. When the French barque *President Felix Faure* ran on to Bollons Island in fog in March 1908, the crew found that the castaway depot, restocked the previous year, had been looted. Instead the men had to survive on seabirds, marine mammals, fish and a calf — the only survivor of a small herd of cattle, sheep and goats placed there by the government steamer six years earlier to sustain castaways.[11]

Steamers had no need to swing far south to catch the winds, so after the last European trade sailing ships were phased out in the early 1900s and the new Panama Canal (1914) provided a new route for shipping, traffic virtually ceased in these inhospitable waters. The sub-Antarctic islands reverted to nature and the Marine Department made its last depot inspection in 1927.[12] In recent times the huts, signposts and other archaeological features have become a scientific curiosity, as the Department of Conservation surveys and records the islands' historic and natural heritage. Many depots have crumbled away, but visitors lucky enough to venture south in modern craft can still see the eerie castaway finger posts, the *Grafton* wreck, the *Derry Castle* and the *Hardwicke* cemeteries, the Enderby Island castaway boatshed, the Erebus Cove castaway boatshed and the unusual A-frame *Stella* Hut provision depot on the Auckland Islands, as well as castaway depots on the Antipodes and the Snares Islands.[13]

OPPOSITE *Hinemoa* crewmen pose beside the frame of the primitive coracle used to get from Disappointment Island to Auckland Island.
Ref: PA1-q-228-09-3, Alexander Turnbull Library, Wellington, NZ

6

'Terrible engines of destruction' — collisions

Victorian journalists loved purple prose. Reporting the funeral of victims of the *Pride of the Yarra* in 1863, the *Otago Daily Times* described the bodies as: 'delicately handled, as sacred though now-deserted temples in which the divine spirit had dwelt; delicately coffined, they were borne to the house of God where the burial service of the church was impressively performed by the ministers of their own faith; and followed by thousands of reverent mourners, amongst whom there was not a dry eye, they were carried to their last resting place in that "God's acre" [the Southern Cemetery] where their remains will never be disturbed by sacrilegious hands'.[1]

The accident alluded to took place in July 1863 and, in terms of human life, it was New Zealand's costliest two-ship collision.[2] The vessels involved were tiny, the harbour steamers *Pride of the Yarra* and *Favourite,* but the deaths of 12 people brought Dunedin to a standstill.

The 25-ton screw steamer *Pride of the Yarra* had already acquired an invidious reputation in four brief years on Otago Harbour. This long, narrow craft — 22.9 metres long and 2.6 metres in beam — had been built in Melbourne in 1856 as a river ferry. In June 1859, just months after entering service on Otago Harbour, the *Pride*'s boilerplates collapsed, dousing its fires while off Grassy Point. Completely helpless, the ferry drifted onto a sandbank where it sank after the tail-shaft pulled out. Only the shallowness of the water prevented heavy loss of life because its owner — that pillar of the business community, William Hunter Reynolds — had not provided even basic lifesaving equipment. The passengers huddled together for warmth on the *Pride*'s upper deck until they were rescued next morning. Although newspapers called for steamers to carry life-saving equipment, the *Pride* still lacked lifebelts or a lifeboat as late as March 1860, by which time two drownings had taken place from the ship — a master and an engineer. A lifebuoy was all there was, stapled to the bridge rail that would have had to be pulled free in an emergency.[3]

That absence of proper lifesaving equipment cost lives on the evening of 6 July 1863 when,

OPPOSITE Six days later and high and (reasonably) dry in the Jubilee floating dock, the huge gap in the *Waipiata*'s hull is obvious. Nonetheless, observers considered that had it not been for the cushioning effect of the packed cargo of wool, hides, bran, blood and bone, and wheat pollard, the bow of the *Taranaki* would have cut the *Waipiata* in two.
Ref: 114/146/10-G, Alexander Turnbull Library, Wellington, NZ

as the *Otago Daily Times* put it, 'two of the port steamers, proceeding at full speed in opposite directions, came into violent collision, and, amid the darkness, the confusion and the general terror which prevailed, the more tender vessel of the two filled and sank, taking down with her many of the human freight and leaving other waifs upon the waters to battle desperately for life — some with success, others with hopeless effort to avert their fearful fate.'[4]

At about 1715 hours the *Pride of the Yarra* embarked passengers at Port Chalmers, some from the jetty, the others from the steamer *William Miskin,* which had just arrived from Invercargill.

Amongst the passengers were nine members of the Campbell family, fresh from London the day before on the migrant ship *Matoaka*. The night was so foggy and dark (and because tickets were usually collected during the voyage) that the *Pride*'s master lost count of how many were aboard; he later guessed 40–50.

Most were up on the chilly deck but some, including most of the women and children, had sought 'the fatal shelter of the cabin'. They included Julia Campbell, wife of the newly appointed principal of the new high school; her two maidservants, Fanny Finch and Mary Roberts; her husband, the Rev. Thomas Campbell; their five children, Edward, Duncan, Muriel, Alfred and Lillian; Elizabeth Anderson, James Frost, Sarah Roberts and Charles Sommerville. Ironically, the Campbells had taken the *Pride of the Yarra* on the suggestion of Captain Dickie,

Samuel Calvert's engraving of the collisions of the *Favorite* and *Pride of the Yarra*, as published in *The Australian News* in late August 1863.
IAN25/08/63/9, State Library of Victoria

PAINTINGS BY ERIC HEATH

These colour pages show paintings of ships, later wrecked, as they were in their prime.

Reproduced courtesy of Eric Heath

ORPHEUS In terms of lives lost, New Zealand's costliest wreck was that of HMS *Orpheus*, lost on 7 February 1863 in the Manukau Harbour at the cost of 189 officers and men, including Commodore William Farquarson Burnett. For the British, this also made it the costliest day of the New Zealand Wars, with the *Orpheus*' mission to New Zealand being to arrange for the withdrawal of Royal Navy sloops *Miranda* and *Harrier*.

LENGTH 69 m
GROSS TONNAGE 2365
BUILT Chatham Dockyard, Kent, 1861

HYDRABAD Running aground during a severe storm on 24 June 1878, the wreck had become something of a local celebrity by the early 1900s – a muse of sketch artists and source of interest for visitors to nearby Foxton, while a walk from Waitarere or Hokio to the wreck was a staple of summer. Indeed, such was the fame of the wreck that when in the early 1930s it was found that the Hydrabad's ornate figurehead had been scavenged by relic hunters, the loss of this 'interesting link with the past' was lamented both locally and by naval authorities, who had been keen to add it to the department's collection at the Auckland naval base.

LENGTH 70 m
GROSS TONNAGE 1350
BUILT Robert Duncan & Company, Port Glasgow, Scotland, 1865

TAUPO On the morning of 18 February 1879, the Union Company's *Taupo* grounded off the Tauranga bar. Accounts of the severity of the grounding differ – some passengers sleeping through it, while others described it as a violent striking against a rock, after which the ship 'appeared to lift under our feet'. The captain immediately gave orders to reverse the engines, but as the vessel began to move the engine room started to fill with water. Passengers and crew were evacuated to safety without fuss and by the next day the *Taupo* was so full of water she would not refloat at high tide. Two years passed before she was successfully refloated and at the end of April 1881 *Taupo* left Tauranga in tow for Auckland but she never made it, springing a leak and sinking in 70 metres of water off Mayor Island.

LENGTH 66 m
GROSS TONNAGE 720
BUILT William Denny and Brothers Ltd, Dumbarton, 1875

WAIRARAPA New Zealand's third-worst shipwreck, the Union Steamship Company's *Wairarapa* ploughed into Great Barrier Island in the first minutes of 29 October 1894, killing 121 (some reports said 125) people. Only weeks before, the *Wairarapa* had set a record for the Sydney to Auckland crossing; on its fateful final trip speed would once again be a factor as the captain steamed along at full speed, despite a thickening fog as the ship moved south from the Three Kings Islands.

LENGTH 87 m
GROSS TONNAGE 1786
BUILT William Denny and Brothers Ltd, Dumbarton, 1882

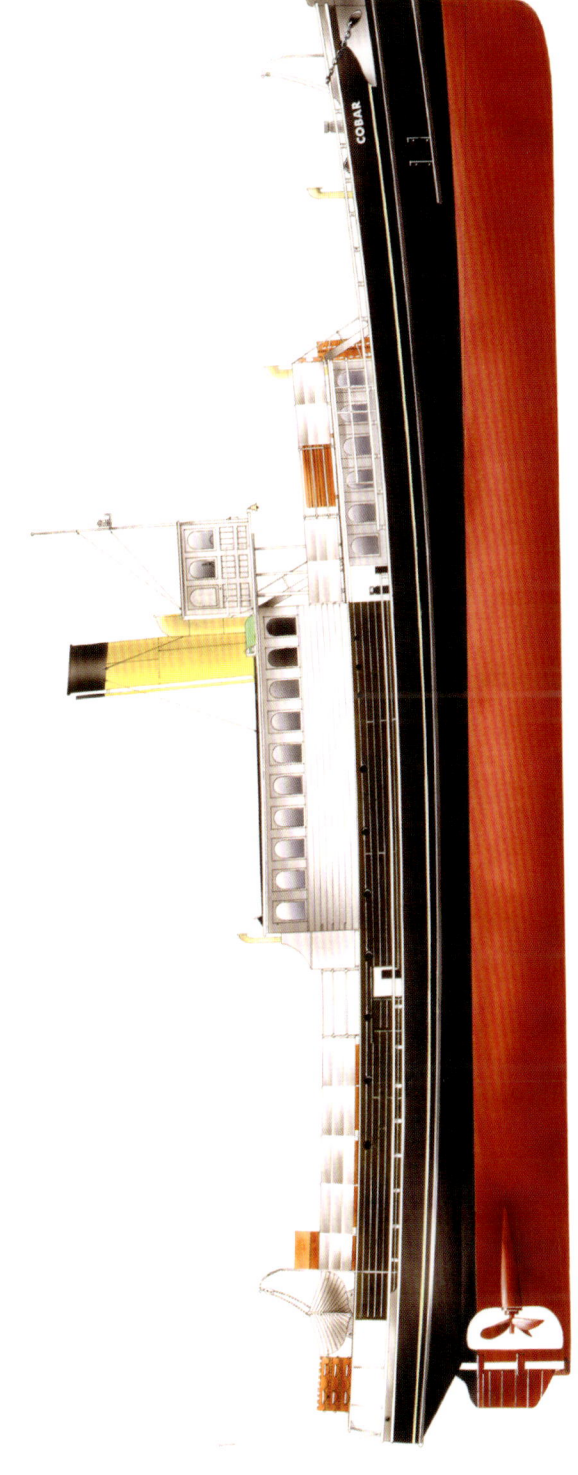

TYRONE The largest vessel to date wrecked on the coast of the South Island, the Tyrone struck on the rocks just south of Otago Heads early in the morning of 27 September 1913. The primary cause was put down to heavy fog, which capping the land, obscured the light on Taiaroa Head, and rendered it impossible to determine the direction of the fog signals that were heard from the lighthouse. The initial grounding was soft, intimated only by lead-line registers and the ship's lack of response to engines full astern. She was assisted to move astern but could not be kept there and settled, broadside, on Wahine Point. Before long the pounding sea pushed her onto the rocks and the hull broke in two. Thankfully, this was not before the cargo of whisky and general merchandise was salvaged.

LENGTH 137 m
GROSS TONNAGE 6664
BUILT Workman, Clark & Company Ltd, Belfast, 1901

COBAR Originally built as a private pleasure craft for the Australian industrial entrepreneur William Longworth, the Cobar was bought by the Wellington Harbour Ferries Company in 1906. Servicing Eastbourne from 1913, the Cobar would be the longest-serving Wellington ferry, plying the service until 1948 when the vessel hulked in the Chatham Islands after being badly damaged by fire.

LENGTH 113 m
GROSS TONNAGE 159
BUILT George de Fraine, Camden Haven, NSW, 1903

RANGI The *Rangi* was one of some 130 scows built in the north of New Zealand between 1873 and 1925. These scows became the transport workhorses of New Zealand's coastal waters and inland waterways. Conceived as a shallow-draught vessel for the on-deck stowing of timber, the scow was designed so that it would remain upright when aground and so allow a timber cargo to be loaded in close to the sawmill, or discharged no matter the tide.

LENGTH 30 m
GROSS TONNAGE 98
BUILT George T. Nicol, Auckland, 1905

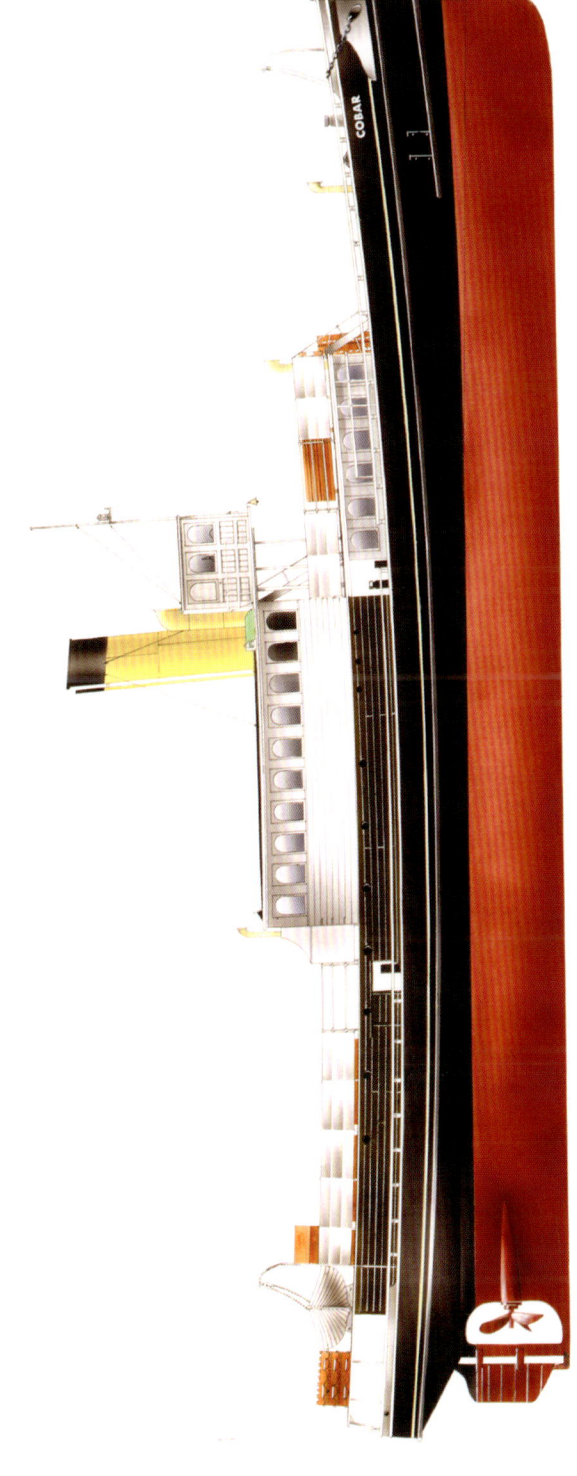

TYRONE The largest vessel to date wrecked on the coast of the South Island, the Tyrone struck on the rocks just south of Otago Heads early in the morning of 27 September 1913. The primary cause was put down to heavy fog, which capping the land, obscured the light on Taiaroa Head, and rendered it impossible to determine the direction of the fog signals that were heard from the lighthouse. The initial grounding was soft, intimated only by lead-line registers and the ship's lack of response to engines full astern. She was assisted to move astern but could not be kept there and settled, broadside, on Wahine Point. Before long the pounding sea pushed her onto the rocks and the hull broke in two. Thankfully, this was not before the cargo of whisky and general merchandise was salvaged.

LENGTH 137 m
GROSS TONNAGE 6664
BUILT Workman, Clark & Company Ltd, Belfast, 1901

COBAR Originally built as a private pleasure craft for the Australian industrial entrepreneur William Longworth, the Cobar was bought by the Wellington Harbour Ferries Company in 1906. Servicing Eastbourne from 1913, the Cobar would be the longest-serving Wellington ferry, plying the service until 1948 when the vessel hulked in the Chatham Islands after being badly damaged by fire.

LENGTH 113 m
GROSS TONNAGE 159
BUILT George de Fraine, Camden Haven, NSW, 1903

RANGI The *Rangi* was one of some 130 scows built in the north of New Zealand between 1873 and 1925. These scows became the transport workhorses of New Zealand's coastal waters and inland waterways. Conceived as a shallow-draught vessel for the on-deck stowing of timber, the scow was designed so that it would remain upright when aground and so allow a timber cargo to be loaded in close to the sawmill, or discharged no matter the tide.

LENGTH 30 m
GROSS TONNAGE 98
BUILT George T. Nicol, Auckland, 1905

WILTSHIRE In one of the most sensational shipwrecks in New Zealand's maritime history, the *Wiltshire* was totally lost on the night of 31 May 1922 at Rosalie Bay, Great Barrier Island. Fortunately, there was no loss of life. Weather was a key factor behind the wrecking: heavy weather and torrential rain, coupled with a dark night, made it impossible to see a ship's length ahead. As a result, the first sign of trouble was the terrific bump as the *Wiltshire* struck, followed by four distinct jumps as the impetus forced her on the rocks. The severity of the weather, however, rendered the crew unable to discern the vessel's fate, which finally became clear around 1130 hours when, with what sounded like a thunderclap, the *Wiltshire* snapped in two abaft No. 4 hatch.

LENGTH 161 m
GROSS TONNAGE 10,390
BUILT John Brown & Co. Ltd, Clydebank, 1912

SOUTH SEA The *South Sea*'s collision with the *Wahine* on the morning of 19 December 1942 was strange, given its occurrence in broad daylight with good visibility. It was the fault of the *South Sea* commander, and the former steam trawler sank in 40 minutes — this delay purely the result of the tug *Toia* and the minesweeper HMNZS *Rata*, who worked their pumps furiously in an effort to keep the *South Sea* afloat long enough to tow her to shallow water. Rather than salvage her, the Navy used the easier and cheaper alternative of stripping vital fittings, removing guns, munitions, radio equipment and other valuables in early 1943. A depth charge was used to blast away the higher parts of her superstructure to minimise the danger to shipping.

LENGTH 43 m
GROSS TONNAGE 322
BUILT Goole Shipbuilding & Repairing Co. Ltd, Yorkshire, 1913

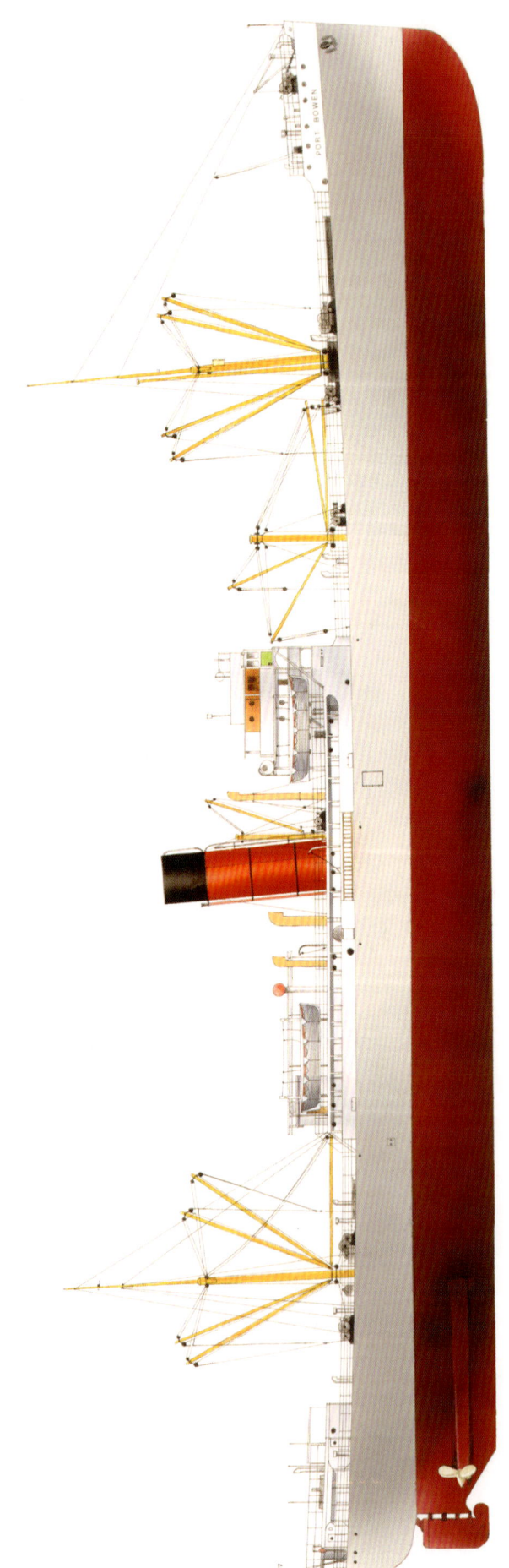

PORT BOWEN Following the failure of efforts to refloat the stranded *Port Bowen* – beached off Castlecliff near Wanganui in July 1939 – its salvage in the context of wartime shortages saw the vessel become almost 'worth her weight in gold', as the *Evening Post* put it in January 1943. Steel from the *Port Bowen* went into the making of huge 80-ton shields employed in driving tunnels during the construction of the Tekapo hydroelectric development. Machinery from the *Bowen* was put to many uses: some went to freezing works, some to hospitals, and some to minesweepers. One generator was even large enough to supply the full electric power needs of the Wanganui Public Hospital.

LENGTH 146 m
GROSS TONNAGE 8267
BUILT Workman, Clarke & Co., Belfast, 1916

the deputy harbourmaster, who said that the *Pride*'s ability to come alongside the *Matoaka* would avoid the danger of transferring the fragile Elizabeth Campbell and her children into small boats. As the night was particularly cold, another small knot of men huddled together in the unlit, partially filled forward hold.

The *Pride of the Yarra*'s regular master, Captain Robert Spence, was at the wheel that night, assisted by an experienced helmsman, James Deuchrass. Seaman James McKinlay was stationed at the bow to keep a look-out, and engineer Charles Perry was below.

After passing through the islands, the *Pride* rounded Kilgour Point, at the start of the upper harbour, where McKinlay spotted the lights of another ship: a white masthead light and green (starboard) and red (port) sidelights. This was the 68-ton paddle steamer *Favourite,* returning from towing a ship up to Dunedin. Captain William Adams and helmsman Charles Murray were on the *Favourite*'s bridge with a port pilot.

The *Pride* was near the middle of the 275-metre wide channel but now swung across to the starboard or right side, following the universal practice for ships to pass 'port to port' with their port (left) sides facing each other. Spence was expecting the *Favourite* to also turn to its right and was alarmed when he continued to see all three of the other ship's lights. This meant only one thing — it was on a collision course. As the paddle steamer surged towards the *Pride of the Yarra,* Captain Spence ordered 'hard a port'. Passengers later also recalled hearing him cry 'For God's sake, reverse the engines!' To his horror, the *Favourite,* still making about seven knots, now turned to port, straight into him. The *Pride*'s bow was swinging away from danger, but nowhere fast enough, even though the *Favourite* had finally stopped its engine. The *Favourite,* 'a terrible engine of destruction', ploughed into the *Pride.* The paddle steamer struck nearly in line with the *Pride*'s mast, cutting right through the ship's port side and listing it over.

'As the water was heard to rush into the vessel's hold and as the deck was felt to subside below the fickle surface, the crowd on deck advanced with all the rapidity which love of life could inspire, to the point of attachment between the two vessels,' a survivor later reported. In fact, it was a mad scramble, because the *Pride,* lacking watertight bulkheads, was sinking fast and the extra weight of desperate people scrambling forward only drove its shattered bow deeper down into the water. A lucky few made it over the *Favourite*'s bulwarks with only their feet wet, but most were left floundering in the chilly water, clinging together and crying out frantically to the crew of the *Favourite,* now scrambling to render assistance.

Not so lucky were those below decks in the *Pride.* The men in the hold up behind the bows, forewarned by a call, made it to safety by the smallest of margins, the first one out being neck-deep in water by the time he started moving. Less fortunate were the saloon passengers. Captain Frederick Wilson from the *William Miskin* and Thomas Kingston, who were seated closest to the door, shot out with seconds to spare, escaping as the water pouring in reached chest-height. Captain Spence pulled out Captain McLellan, the lame skipper of the cutter *Alpha,* but could reach no one else before his vessel slipped beneath the waves. As for the Campbell family, 'so happy in the knowledge of arrival at their new home ... cribbed, cabined and confined, they had not even the drowning man's hope'.

The *Pride* filled fast, lurched once and then sank, taking with it the dead and the living. Because the *Favourite* did not carry a boat, the paddle steamer could do little to help the people floundering in the icy water: 'round the stricken steamer a fearful arrival of death was enacting, and the scatheless victor in the fearful encounter had to stand supinely by', the editorial writer lamented a couple of days later. One man floated free and was hailed by the *Favourite* but was too weak to grasp the line thrown to him. He uttered a pitiful gasp, then sank from view. The

Favourite loitered for half an hour, then steamed for Port Chalmers where the shaken passengers transferred to the *Golden Age* that took them to Dunedin.

Next day harbour staff and the police used divers to recover bodies from the wreck. The divers reported finding the bodies of the children gathered around the mother, Julia Campbell clutching her infant with one hand and her husband's with her other. The diver carefully wrapped the infant's body with a shawl before sending it up to the surface. The tragedy stunned Dunedin, which had been looking forward to the Campbells' arrival for a long time. The public funeral, which followed on Thursday, brought the town to a standstill. Two thousand mourners, almost the complete population, attended and the procession stretched for almost two kilometres.

The inquiry convicted the master and mate of the *Favourite* of manslaughter and severely censured Captain Spence for driving his ship at excessive speed. A recent book on the collision offers some more support for Captain Adams' assertions that the wind blowing the *Pride*'s funnel smoke obscured that craft from the *Favourite*'s lookout and that the *Pride*'s navigation lights were inadequate at the best of times.[5]

The harbourmaster tightened up safety procedures and larger, better-equipped ferries replaced the pioneering craft in the next couple of years. The *Pride of the Yarra* was raised and beached but never sailed again. When a survey revealed the extent of the damage, it was condemned and left to deteriorate on the shoreline. In July 2003 a plaque was unveiled to commemorate Dunedin's 'harbour horror'.

Northern collisions

Minor collisions were common a hundred years ago or more as large numbers of small coasters shuttled in and out of busy waterways. In most cases it was no more than a snapped bowsprit, some battered bulwarks or a few stove planks. Even when ships sank, the shallow water often made salvage easy.

Auckland's Northern Steam Ship Company's fleet was typical. Formed in 1881 from a number of concerns controlled by Alexander McGregor, its small steamers serviced the Northland, Hauraki Gulf and Bay of Plenty outports the Union Company ignored.

McGregor had already come a cropper in 1879 when his *Glenelg* challenged the rival *Rose Casey* to a race out of Auckland. It sprinted after its rival and had almost caught up when the German warship *Albatross,* lying peacefully at anchor in the stream, loomed into view. The *Rose Casey* was giving no ground, so the *Glenelg* smashed into the warship's gig, which was off its stern at the time. McGregor had to apologise and offer the Germans a replacement boat.

Racing was one of the great unpardonable sins and officially, at least, shipping lines took a dim view of it. It did happen, though, and it led to the occasional accident. In February 1888 the Northern Company's *Gairloch* — trying to beat the Seamen's Union's *Bellinger* into Waitara — hit the north training wall; days later the same company's *Clansman* ran down the cutter *Mana* off Mercury Island.

Northern's worst collision involved two of its steamers, the *Wellington* and the *MacGregor*. It took place off Kawau Island on the night of 11 May 1885. The *Wellington* was steaming at 11 knots when the *MacGregor*'s white masthead

OPPOSITE, TOP Five excursionists drowned when the *Claymore* ran down the *Kapanui* in 1905. This artist's impression captures the horror.
Gavin McLean Collection

BELOW Despite the sizable hole the *Claymore* had punched in the side of the *Kapanui*, the damaged vessel was ultimately salvaged. Four years later she was not so lucky, being totally destroyed by fire at Warkworth Wharf, Auckland.
Gavin McLean Collection

91

light was spotted off the bow. Chief officer Stephenson starboarded his helm and called the master to the bridge, but the notoriously 'cranky' *Wellington* responded sluggishly and hit the *MacGregor* amidships, nearly cutting it in two and sinking it. The only casualty was the *MacGregor*'s second officer who was seriously injured by falling timber.

It took nine weeks, £3700 and a great deal of ingenuity to salvage the *MacGregor*. The first lifting attempt failed when one of the chains wrapped under the hull broke. On 29 May the ketch *Fanny Thornton* and the cutter *Gazelle* succeeded with these chains but at great cost to the elderly *Gazelle,* which sank and had to be salvaged. The *MacGregor* finally made it back to Auckland on 15 July.

An excursion turned to disaster on 23 December 1905 when the steamer *Claymore* ran down the *Kapanui* off Devonport Wharf, ripping open the smaller ships' port side and drowning five passengers. As the water was shallow, the scows *Kauri* and *Rata* were positioned either side of the wreck, spaced apart by huge baulks of timber secured across each scow. Wires were then passed under the *Kapanui* and attached to the lifting vessels. At high tide the wires raised the *Kapanui* off the seabed, enabling the steamer to be moved into shallower water. After several such moves it was beached for repairs.

Collisions on Wellington Harbour

One of the strangest collisions took place in broad daylight on Wellington Harbour during wartime. The ships involved were the Union Company's inter-island ferry *Wahine* and the auxiliary minesweeper HMNZS *South Sea*. The *Wahine* had been entering and leaving Wellington Harbour since 1913. Its master, Captain Alexander Howie, knew the port very well. The *South Sea* (312 tons), commanded by Temporary Lieutenant Peter Bradley, was the former steam trawler *Ferriby,* built in England in 1912. Requisitioned by the navy as an anti-submarine minesweeping trawler, it was patrolling inside the harbour at the time of its loss.

The weather was fine with good visibility as the *Wahine* pulled away from Fryatt Quay at 0824 hours on Saturday 19 December 1942 and put on a good turn of speed. Six minutes later, just abreast of Point Jerningham, Howie became worried that the minesweeper, heading in from its picket line off Somes Island, looked to be on a collision course. Bradley had seen the ferry but overestimated his own speed and thought that he had time to clear it safely. At 0831 Howie altered course to starboard and sounded the whistle. Another minute later, with the *South Sea* now moving at about 6 knots and sheering slightly to port, he put *Wahine*'s engines full astern. This had cut its speed to about 8 knots by the time that the ship sliced into the small minesweeper two minutes later; the impact was still considerable, however, throwing two men off the stern of the *South Sea* into the water and knocking out the warship's helmsman. The ferry had struck the minesweeper just aft of its bridge, opening a big hole in the starboard side.

Rescue craft appeared on the scene within minutes. The *Portland* plucked the two ratings from the water and the rest were able to step across to other craft without difficulty. The tug *Toia* and HMNZS *Rata,* their pumps working furiously, tried to keep the *South Sea* afloat long enough to tow it into shallow water, but it sank off Point Halswell at 0920 in about 24 metres of water.

The *Wahine*, after careful examination, continued on to Lyttelton. Vital to the ferry run, the ship sailed with a damaged bow until mid-January 1943 when it was repaired at Lyttelton. Although it considered salvaging the old minesweeper, the navy settled for the cheaper alternative of stripping the wreck of its vital fittings. Between January and March 1943, it removed the guns, depth charges, radio gear and other valuable items. It then lowered an old depth charge and blasted away the higher parts of the *South Sea*'s superstructure to minimise the danger to shipping. The old trawler still rests there today, surprisingly intact.

Because of the wartime need for security and the fact that a court of inquiry had no jurisdiction over a naval vessel, there was no formal marine court of inquiry, though the navy held its own inquiry in December 1942. It found that Bradley had erred in not taking any bearings of the *Wahine* and that he had not acted in accordance with articles 22 and 23 of the Regulations for Preventing Collisions at Sea. Howie might have avoided the collision by turning to port rather than starboard (i.e. by turning towards the minesweeper), but this would have been against the regulations and all the instincts of seamanship, since the Union Company master did not know that the warship captain had not realised that a collision was impending. The company later billed the navy for almost £5000.

"Feast for a Number of Eminent Solicitors" — the *Waipiata/Taranaki* collision

Wellington featured again less than a decade later when Shaw Savill & Albion's big refrigerated cargo liner *Taranaki* rammed the Union Company freighter *Waipiata* off Seatoun, just inside the heads, on 5 May 1950. No one was injured but both ships suffered extensive damage.

It was a cold, wintry day with a strong southerly sweeping in patches of heavy, misty rain. The *Waipiata*, a big 2847-ton, engines-aft coaster, was inbound from Oamaru, and the *Taranaki* (8695 tons, 1928) was heading out, bound for Hamburg and London with six passengers and a large cargo.

Wartime regulations prevented photographers from recording the *Wahine/South Sea* collision. This image shows the ferry after an earlier accident when it struck the wharf at Wellington while berthing in heavy fog in June 1936.
Ref: 1/2-147745-F, Alexander Turnbull Library, Wellington, NZ

Both ships were in the hands of experienced men. The *Taranaki*'s Captain H.J. Bennett was one of the last seagoing masters with a square-rigged vessel's qualifications on his ticket. He had commanded submarines in World War I and had been with Shaw Savill since 1932, apart from war service. The *Waipiata*'s Captain James MacNeil had been with the Union Company for 25 years.

The bigger ship slammed into the *Waipiata* immediately forward of the bridge, smashing its starboard wing to pieces. A seaman rushing up on deck was greeted by 'a frightening view of a ship's bow towering over our bridge, cloaked in steam from our fractured pipes'.[6] Captain MacNeil swung out the lifeboats while the *Taranaki* rigged nets and floodlights in case the *Waipiata*, which was settling by the bow, sank.

The *Taranaki*'s passengers, seated at dinner, had felt the impact as little more than a sudden roll. In the much smaller *Waipiata* it felt far worse. The 31 crew put on lifebelts. Calmest of all was Mary, the *Waipiata*'s greyflecked cat. When they feared that their ship might sink, some seamen tried to put her aboard the pilot launch, but she fought so stubbornly that they gave up. When put ashore the next day she returned promptly, as if to say, 'after all the crew hasn't been paid off yet'.

Fortunately the *Waipiata* stabilised with her foredeck awash. The pilot boat crew tried and failed to prise apart the two ships, then gave up, deciding that the undignified embrace was keeping the coaster afloat. But the ship was a mess. The forepeak, chain locker and No. 1 and No. 2 holds flooded as it settled in the water with less than a metre of freeboard at the No. 1 hold. Fortunately, crates of acid stowed in the deck cargo remained intact. Had they burst, the crew might have been unable to keep the ship afloat.

As the *Waipiata* showed no signs of settling any further, harbourmaster Captain D.M. Todd decided to bring in both ships 'as they lay'. And so, with the *Waipiata* hanging across its bows, the *Taranaki* crawled up harbour stern-first. Sight seekers swarmed along the Oriental Bay waterfront in such numbers that a traffic officer was needed there to control traffic. The trip to the Clyde Quay Wharf took about five hours, finishing at midnight. In the final manoeuvring, the *Waipiata* provided some assistance by using its engine. Slowly and tenderly, the ship was edged around until it lay alongside the wharf in the normal fashion, its sunken bows pointing out to sea.

There the tugs *Taioma* and *Tapuhi* tried again to separate the ships but merely broke mooring lines, bent some of the *Waipiata*'s forecastle

The *Taranaki*'s bow looms over the *Waipiata* in the dark. The tin barrels visible on the smaller *Waipiata*'s deck are thankfully linseed oil as opposed to something more explosive.
Ref: 114/100/14-G, Alexander Turnbull Library, Wellington, NZ

railings and raised everyone's blood pressure. Eventually, workmen using oxyacetylene gear made some impression and the *Taranaki* and *Waipiata* were separated at about 0800 on the 6th. During this dangerous task, paint in the *Taranaki*'s bow locker caught fire, but the fire brigade quickly extinguished it. Once freed, the *Waipiata* settled bow-first on the bottom alongside the wharf, its forward holds by now an unappetising mush of cereals, peas and Bournvita.

The coaster had come off much the worse, being sliced in two almost to the keel. The No. 2 hold had been penetrated to a depth of at least two metres and flooding had affected both holds as well as the forepeak. The *Taranaki* had also smashed in the second officer's cabin, which lay beneath the starboard bridge cab. Had second officer Athol Hansen been asleep at the time, he would probably have died. The *Taranaki* had taken a little water through its damaged bow but most of the damage had occurred above the waterline. The ship entered the floating dock in June 1950 and reappeared four months later with a shiny new stem. The *Waipiata*'s repairs took much longer.

The court of inquiry, held at Wellington in June under the chairmanship of A.A. McLachlan, SM, heard much conflicting evidence. It found that both masters were unusually tardy in acting under the provisions of the Regulations for the Prevention of Collisions at Sea and apportioned the blame three-quarters to MacNeil and one quarter to Bennett. Had MacNeil applied ordinary cautionary measures eight minutes earlier, by moving to starboard in accordance with the 'end-on rule', collision would have been avoided; had either done so even four minutes later, slowing up, stopping or going astern, no damage would have been done. MacNeil had probably mistaken a green section of the Somes Island light for the *Taranaki*'s green light and had compounded his error by pressing on even after he suspected that something was wrong. When the court announced that it would restore their certificates, Captain Bennett crossed the court room to shake MacNeil's hand.[7]

The *Taranaki*'s damage was almost superficial in comparison with the *Waipiata*'s. The big freighter built by Fairfield Shipbuilding at Govan (near Glasgow) was a lucky ship, surviving several collisions, the worst being the sinking of the coaster *Lairdsmoor* near the UK in 1933, and coming through World War II unscathed. By the time the ship was delivered to Japanese ship breakers at Aioi in September 1963, it had steamed 1,875,799 nautical miles and passed through the Panama Canal 74 times and the Suez Canal 22 times. Napier people still remember the *Taranaki*, which was lying in the roadstead in February 1931 when the earthquake struck, thrusting up the seabed. The *Taranaki* was lucky to scrape across the seabed to deeper water two nautical miles further out. The crew helped rescue survivors and three days later took the first party of refugees to Auckland. The other ships in the class were the *Coptic*, *Zealandic* and *Karamea*.
Ref: 114/100/09G, Alexander Turnbull Library, Wellington, NZ

'THAT LONG DRAWN OUT TORTURE' — THE INQUIRY

Last century the Marine Department oversaw the investigation of marine casualties. Not every accident warranted the cost and fuss of an inquiry. If the accident was trivial or the cause was obvious, an inquiry was not needed. Sometimes this meant that major incidents — such as the mining of the liner *Niagara* in 1940 — where the cause, a German mine, was well known, were not investigated by a court.

Marine Department officials often disagreed privately with court judgments and the *Waipiata/Taranaki* collision was a good example. In July the department's nautical adviser vented his spleen. The court's report, he advised the departmental secretary, was wordy and hard to follow. The hearing had taken six days when it should have needed only three. He disagreed with its assertion that Captain Bennett's decision to take the *Taranaki* out without a pilot was 'bold' — Wellington is not difficult and is well marked, he snorted.[8]

He was far more concerned by the court's acceptance that a local rule existed in Wellington. 'The inference re the existence of a local rule is contrary to fact, no local rule as stated has ever been accepted,' he wrote, 'and it would appear that Counsel for Captain MacNeil sold the idea to the court and to all other Counsel.' Captain Bennet 'had no need to consider even for a moment that a rule which radically infringed the Regulations for Preventing Collisions at Sea would have the approval of the Marine Department and the Harbour Authority.'

'I do not agree with the finding in so far as the proportion of blame is concerned,' the official advised, 'but it would be futile to re-open the case if we had to witness another exhibition of that long drawn out torture which was inflicted on the unfortunate Captains involved in this Marine Casualty.'[9]

The secretary agreed and advised the minister accordingly. 'The case certainly became a "feast" for a number of eminent solicitors and the finding of the court was a strange one,' he told the politician.[10] The department had already arranged to meet harbour board representatives to ensure that there were no more misunderstandings about 'local rules'.

The *Waipiata* alongside Clyde Quay Wharf (now the site of the Overseas Passenger Terminal) the following day.
Ref: 114/100/03-G, Alexander Turnbull Library, Wellington, NZ

"COSPATRICK" CAPE OF GOOD HOPE BURNT 19 NOV 1874. 680 LIVES LOST.

7

'A thrill of horror throughout the Empire' — fire

In 1876 Jane Findlayson, a passenger on the immigrant ship *Oamaru,* bound for a new life in New Zealand, stood on the ship's deck and 'looked out across the cold grey sea to the spot where the *Cospatrick* had caught fire'.[1] One author of a book on life aboard the immigrant ships observed that this brief diary entry showed how deeply the disaster off the African coast had affected immigration. The burning of this ship was blamed for a sharp decline in immigrant numbers after 1874. It also changed shipboard behaviour and officials' attitudes. Before the *Cospatrick* disaster, the government's immigration commissioners usually did not ask shipping companies about their fire drills. After 1874 they frequently did and sometimes demanded demonstrations'.[2]

Of all the perils nineteenth-century ships faced, fire, especially far out at sea, was feared the most. All too quickly, fire could quickly turn small, crowded wooden ships, lit only by candles and oil lamps, into death traps. Fire-fighting equipment was rudimentary — the breathing apparatus needed to fight below-decks blazes simply did not exist. Another problem was cargo stowage — it was common to scatter petroleum products, paint and other highly combustible products among general cargo. Because of the danger, few lights were left burning at night untended, adding to the gloom of the passage. On the better immigrant ships, late-night fire patrol duties were added to the crew and passengers' shipboard tasks.

The blackest spectre that haunted nineteenth-century imaginations was that of the New Zealand immigrant ship *Cospatrick* burned at sea on 17 November 1874, with the loss of almost 470 lives. This disaster symbolised the worst fate that the sea had in store. In the words of veteran shipping journalist Sir Henry Brett, it 'caused a thrill of horror throughout the Empire and particularly in Auckland, for which the vessel was bound'.[3]

The *Cospatrick* was a two-decked full-rigged ship of 1220 tons built to the now outmoded Blackwall Frigate style in 1857. The 1874 voyage was just its second to New Zealand after Shaw Savill had bought the ship in 1873.[4] In command was the 39-year old Alexander Elmslie, who had with him his wife Henrietta and their son Alexander. On board

OPPOSITE The *Cospatrick* at Port Chalmers. The ship's old-fashioned, bluff lines are obvious.
Ref: 1/2-012556-a-G, Alexander Turnbull Library, Wellington, NZ

were 479 people — 44 crew and 435 passengers. If that does not seem crowded, remember that the 65,000-ton liner *Queen Elizabeth II* was designed to carry 1400 passengers.

Most of the passengers, government-assisted immigrants, would never see their destination. The *Cospatrick* was several hundred sea miles west of Cape of Good Hope when fire broke out in the forepart of the ship, probably, it was later decided, by crew pilfering stores there. Boatswain Henry McDonald, one of the very few survivors, recalled being awakened from his slumbers by the dreaded cry 'Fire!' Although unclothed, McDonald ran forward to find great clouds of smoke issuing from the forepeak. The bosun's locker, crammed full of flammables such as oakum, rope, varnish and paint, was well alight. The crew got the hoses going, spraying the fore peak with seawater, but made no impression on the flames. Soon the *Cospatrick* lost steerageway and — its head to the wind due to an error by a ship's officer in raising the foresail above the fore-hatch — burned from stem to stern as the breeze fanned the flames aft. Within an hour and a half of the discovery of the first wisps of smoke, the *Cospatrick* looked doomed.

According to McDonald, 'dreadful scenes followed, for a panic broke out among the emigrants'. Some rushed for the boats, which exceeded regulations by having space for 183 people — less than half those aboard the *Cospatrick*.[5] Two of the six had already been consumed by the flames. The first boat launched was immediately swamped by the crowd of 'demented men and women that jumped into it'.[6] The longboat went up in flames and only two boats were launched safely.

The occupants of the boats had to watch the last moments of the *Cospatrick*, powerless to assist their friends. As McDonald recalled: 'The main and mizzen masts fell, and many of those who had crowded aft were crushed to death. Then the stern was blown out. That was the end, and the shrieks of the survivors were silenced suddenly in the roaring flames.'[7] Yet, as he later admitted, the ones who died then were perhaps the lucky ones.

The people in the boats had insufficient clothes, food or sails and were hundreds of miles from land, entirely at the mercy of the wind and currents. On the 20th a gale separated the boats, one of which was never seen again. Yet the people in the remaining boat had little energy or inclination to mourn the dead. Without food or water, they began to succumb one by one.

Their desperation eventually drove some to observe 'the custom of the sea', cannibalism, slicing open corpses to eat their livers and drink their blood. Their ranks thinned with every day. A man, steering, fell overboard and drowned. Three went mad through lack of water and died. Their bodies were thrown overboard. Four more died on the 24th, three of them raving madmen. Early the next morning, a steamer passed within 50 metres without spotting the boat; another man died that day. Two more the next.

Although the few survivors threw one body overboard, they were now too weak to get rid of the other corpse, which they had to leave rotting in the bottom of the boat. Just five men remained alive and drinking seawater was fast threatening their remaining shreds of sanity. Then, the *British Sceptre* put in a timely appearance. Their rescuers

The *Cospatrick* burning at sea — a contemporary reconstruction of the event.
Gavin McLean Collection

shuddered at the semi-devoured corpse and the appalling state of the barely-living: 'their starved-looking faces and emaciated bodies ... [which] too plainly revealed the dreadful sufferings that had been endured'.[8]

Harbour blazes

Even at anchor, safe and sound in port, wooden sailing ships were vulnerable to fire. The migrant ship *Montmorency* caught fire in the Napier roadstead in March 1867. A 688-tonner dating from 1855, this former Black Ball liner was described as 'remarkably roomy between decks', and was a favourite vessel for the conveyance of immigrants — allegedly it had 'carried more immigrants in her day than any other British ship'. Since reaching Napier on the 24th it had discharged its passengers and their luggage, but had not started on the cargo of 400 tons of general merchandise for Napier merchants.

Just minutes before midnight on 27 March a watchman discovered smoke and sounded the alarm. The chief officer (Captain Josiah Hudson MacKenzie, was ashore at the time) and the crew scrambled up on deck. As soon as he went forward, the chief officer found his way barred by thick smoke billowing out through the hatchway. He ordered his men to pass down a hose and to prise apart the hatch cover; this revealed flames coming from the port side. The chief officer, boatswain and sail maker went down into the hold, playing the hose on the fire, but the thick, choking smoke and the heat soon forced them back. They closed up the hatch cover to deprive the fire of as much oxygen as possible and warned the crew to prepare to abandon ship. While they were doing this, the chief went aft to burn blue lights and fire rockets to attract attention.

Within minutes the flames had spread from the hatches to the spare spars stacked on the decks. The *Montmorency*'s position was now hopeless. Seamen who tried to go below to fetch their possessions were forced back by the flames; the ship's carpenter's efforts to scuttle the vessel were foiled by the same smoke and flames, and by the swell. By 0030 the flames were licking the poop. More rockets burst over the anchorage but drew no response. The chief officer then ordered the crew to take to the boats, leaving just the second officer, the port customs officer and himself aboard.

By now, as the *Hawke's Bay Herald* of 30 March recorded, the flames had been noticed by an officer of the barracks:

> The cargo included over 100 casks of spirits, and a quantity of other goods of an inflammable character, and the volume of fire which shot up to heaven was immense, as well as singular in appearance. This sight is described as grand in the extreme, although of course, such as to create a feeling of deep sorrow for such a calamity. After the fire reached the deck the fore-rigging was the first to be ignited; then the main yard, causing the main top sail to drop down on the poop. Soon after, the chief officer and the two others who remained on board were forced to drop over the stern into the boats'.[9]

The boat, accompanied by another commandeered by the port pilot, again tried to scuttle the *Montmorency* but again failed. When they tried to unshackle the ship from its moorings, they freed only one cable, being compelled to abandon their attempt and pull away to watch the masts fall. They splashed overboard about 0430; the foremast went first, over the starboard bow, to be followed, in quick succession, by the others.

All day the wreck smouldered while crowds gathered on the hill to watch. At one point the *Star of the South* tried to unshackle the remaining mooring and tow the *Montmorency* away from the anchorage but failed and suffered minor damage. By the next day the charred remains of the *Montmorency* lay on the beach. The cause of the fire was not established; some thought that it may have been smouldering for a long time at sea, but this seems unlikely because an inspection the day before had produced no sign of trouble.

Steel hulls and more scientific cargo stowage practices reduced the number of ships lost by fire. Even so, accidents still happened, thanks largely to increasing carriage of motor spirits. Modern-style coastal tankers really only entered service locally in the 1950s. Until then, motor spirits, carried in the tins that inventive Kiwis used for dozens of household purposes, formed part of most ships' cargo.

One victim of such cargo was the coaster *Moa*.

The unusual sight of the coaster *Moa* burning near the entrance to the Whanganui River, in early February 1914. Her cargo — consisting of some 5000 cases of kerosene, motor spirits, benzine, and turpentine, and 25 tons of general cargo — exploded without the slightest warning, killing the ship's fireman William Kennedy, who was sitting on the after hatch, instantly.
Auckland Libraries Heritage Collections, AWNS-19140212-39-1

Built at England in 1864 and altered several times during its career, the 185-ton steamer was approaching the entrance to Whanganui at 1000 hours on 3 February 1914 when an explosion rocked the vessel. Within minutes it was alight from stem to stern — not surprising given that the cargo included over 5000 cases of kerosene, motor spirit, benzine and turpentine.

The flames raced along the deck so quickly that they consumed the lifeboats before the crew could get near them. They threw lifebelts and pieces of wood in the water and leaped for their lives, swimming towards the *Arapawa*. So strong was the heat that the other ship, also carrying benzine, had to keep its distance, though the *Arapawa*'s boats picked up everyone from the *Moa* apart from the fireman, William Kennedy, who had died in the initial blast. As for the fiery torch that was once the *Moa*, the winds took hold of it and blew it ashore a few kilometres south of the southern breakwater. There it burned all day until about 2000 hours when a huge explosion put out the flames.

Four years later another little coaster full of benzine almost set the city of Wellington alight. This was the Westland Shipping Company's 185-ton *Defender*. On 2 August 1918 the ship, which had been on the coast for almost 15 years, was moved from Queen's Wharf to King's Wharf where it started to trans-ship 1200 cases of benzine to a larger vessel. At about 1100 hours this work came to an abrupt halt when someone noticed smoke coming from the forehold. When the crew's fire-fighting efforts failed, the harbourmaster, aware that he had a potential floating time bomb on his hands, ordered the tug *Karaka* to tow the *Defender* away from the wharves. 'So far the flames, though raging under the battened hatches, had not been visible above the deck,' the *Evening Post* recorded, 'but almost immediately after the vessel had been towed clear of the larger vessel and the wharf a few minutes after noon, a dense volume of black smoke, followed by a rush of roaring flames, burst from the fore hatch and rapidly increased.'[10]

While city workers held their breath and

motorists and cyclists scampered out onto the Hutt Road to watch, the *Karaka,* escorted by the tug *Terawhiti* and launch *Uta,* towed the blazing hulk across the harbour. Just off Lepers' Island, near the northern tip of Somes Island, the *Karaka* slipped the tow and let the *Defender* drift onto the rocks. At 1325 a huge explosion rocked the city; another blast followed 15 minutes later. The fire was left to burn itself out. Fifteen days later the crew of the steamer *Aorere* reported that only a davit and a length of blackened superstructure poked above the water. An effort was being mounted to recover the machinery.

Lit up at Lyttelton

Twentieth-century innovations in ship design (steel hulls, electric lights and, more recently, smoke detectors) greatly diminished — although did not eliminate — the risk of shipboard fire. In recent times fires have claimed few trading vessels, although many lives have been lost through these blazes.

In the 1950s two modern coasters suffered serious fire damage. The first, Richardsons' *Pateke* (785 tons), caught fire at Lyttelton on 10 December 1955. By the time the local fire brigade — assisted by crew from the US icebreaker *Edisto* — had brought the fire in the accommodation block under control, £326,000 worth of damage had been done to the year-old ship and three sleeping seamen had been lucky to escape. 'The fierce heat made the task of the brigade difficult,' the *Press* reported. 'In places the steel deck was almost too hot to stand on and it had to be cooled with water from the hoses.'[11] Ship's cook Ronald Kilminster suffered burns and shock, and AB Ivor Pearce and OS Bruce Smith shock and smoke inhalation. It was believed that the blaze may have been caused by a cigarette butt tossed through the open porthole in the second mate's cabin.[12]

Four years later another new coaster went up, this time with fatal consequences. Like the

Of all the perils nineteenth-century ships faced, fire, especially far out at sea, was feared the most. All too quickly, fire could quickly turn small, crowded wooden ships, lit only by candles and oil lamps, into death traps.

Pateke, it was a modern (1957), well-appointed ship of the very latest design — the Holm Shipping Company's 845-ton flagship *Holmburn* — and the place was again Lyttelton. The *Holmburn* had been loading general cargo for Lyttelton throughout the afternoon of 7 May. When sailing was put back a day, several crewmen went to the pub. At about 2300 hours the assistant steward and the cook came back to the ship. Less than an hour later, the former detected smoke. The ship's accommodation block was burning. It was never established exactly where the fire started — the starboard half of the upper deck and or the main deck round about the stairways was the court of inquiry's guess — but the strong southerly wind blowing in through an alleyway on the starboard side of the upper deck fanned the flames into the master's and the chief officer's quarters and up to the wheelhouse.

The assistant steward and the cook woke their shipmates, who scrambled ashore as fast as they could. Two men did not make it ashore, Captain Derek Crabtree, and the chief steward, A. J. Hempstalk. The mate, Keith Baker, made a determined effort to rescue them, but it was seen that they were dead. It was later revealed that

The coaster *Holmburn* aground on a sandbank in the Manukau Harbour in the early 1970s.
J.F. Holm Collection

Captain Crabtree had been asphyxiated in his bathroom (the court said that 'further and more persistent attempts should have made to rescue him', as he may have been still alive when glimpsed through his window), and Hempstalk had died through a combination of asphyxiation and burns.

By the time the fire brigade arrived, the *Holmburn* was well ablaze and smoke was billowing across the harbour. One firefighter was injured while bringing the fire under control, but by 0400 hours on 8 May it was all over. At one stage the cook returned to the *Holmburn*'s boat deck but was judged 'intoxicated and troublesome' by a constable and was forcibly removed.[13]

It had been a close thing. The *Holmburn* was listing to starboard against No. 7 East Wharf, its accommodation area battered and blackened. Repairs took months. The court of inquiry ruled out electrical faults, spontaneous combustion or repair work as causes of the fire. It accepted that the *Holmburn* was well designed and maintained and had firefighting equipment that exceeded statutory requirements. There was some doubt about the safety of the plywood linings used in modern ships, but the principal problem was the incompetence of the crew, who (apart from Keith Baker) 'made no use of the vessel's firefighting facilities or of their training in firefighting ... Their conduct is to be deplored.'[14]

The Marine Department convened a meeting between the seagoing unions and shipowners. After the *Pateke* blaze the seamen had threatened industrial action unless changes were made, but had, department officials noted, lost their zeal when it was pointed out that some of the layout arrangements they criticised had been made to suit their members. This time, however, there was broader agreement: changes were made to egress provisions in the rebuilt *Holmburn* and more attention was paid to firefighting training and to the *Pateke* court's recommendation for ships keeping a 'watchful wake' while berthed in port.[15]

'My Children were Washed away One by One' — the *Capitaine Bougainville*

'The lifeboat was turned over ten or a dozen times before we drifted near the beach,' Captain Jeane Raymond Thomas told reporters. 'Each time someone lost their grip. My children were washed away one by one. I was supporting my wife on the keel but at about daylight she died from the cold and exposure and I lost my grip on her.'[16]

Thomas had commanded the French islands trader *Capitaine Bougainville*. The 3614-ton ship had been off the Northland coast in the early hours of 3 September 1975 — the first stage of a routine voyage back to the Solomons via Sydney and Port Moresby — when fire was reported in the engine room. Aboard were 37 men, women and children. It was an old ship, dating from 1955 but, like all Sofrana-Uniline ships, well maintained.

The crew fought a losing battle with the fire, but as the engines still worked, Captain Thomas kept going, heading close to the shore where he hoped to anchor and await daylight and rescue. Later that day he anchored two miles off Whananaki, near the entrance to Whangarei Harbour. But by then, the smoke was so dense that there was no alternative to abandoning ship. 'I realise [sic] that the life boats had to be lowered quickly for their position near the ventilations [sic] inlets and outlets exposed them directly,' he recalled.[17]

Abandoning ship under those conditions must have been terrifying. Thomas estimated that the swell was about five metres high and the wind was howling at around 40 knots. 'The ship rolled and pitch [sic] violently, making this operation very long and difficult ... and the heat was intense. After having made head count and satisfied that nobody else was left on board I left myself the *Capitaine Bougainville*.'[18]

The heavy sea running made conditions difficult in the water. Both lifeboats and one life raft capsized; some scrambled back aboard but 16 perished, either drowning in the icy waters or succumbing to exposure. A Northland Harbour Board tug took the still-burning ship, by now a constructive total loss, in tow and had it alongside the wharf by 1530 hours. There the fire was finally put out.

The incident captured the headlines, and there was some ill-informed armchair criticism of search and rescue and the time taken to fit out the tug *Waitangi* for a sea rescue. The inquest into the deaths of the ten people whose bodies had not been recovered absolved Captain Thomas of any blame for the incident, which was attributed to a spare piston head breaking loose and fracturing a valve from which diesel oil later escaped and splashed over a hot exhaust pipe.[19] The Ministry of Transport's senior nautical surveyor was not so sure. 'The cause of the fire has not been definitely established,' he advised and expressed doubts about whether proper use had been made of the firefighting equipment (particularly the CO_2 system) and whether the ship had been abandoned prematurely.[20] Nevertheless, the department decided against holding a formal inquiry.[21]

The abandoned *Capitaine Bougainville* on fire.
Warren Spiers, *Northern Advocate*

GOTHIC HORROR

In 1969 the Shaw Savill & Albion liner *Gothic* caught fire. Built in 1948, the 15,911-ton *Gothic* had become famous for serving as the stand-in royal yacht in 1953. Although de-rated for cargo-only operations by the late 1960s, it was still an impressive ship. On 2 August, while deep into the Pacific on a voyage from Bluff to Liverpool via Panama, and carrying refrigerated and general cargo, the *Gothic*'s smoke alarms began to ring. A serious fire was raging in the smoke room and adjoining galley. Probably started by a dropped cigarette butt, it spread once the cold sea spray blew in the overheated windows. Unfortunately, the *Gothic* was heading into the wind; this pushed the fire back through the former passenger accommodation in the direction of No. 3 hold, packed with wool.

Captain Agnew turned around his battered command and helped his crew control the flames. By the time that happened, three hours later, seven people, including two young boys travelling with their father, had died. With much of its superstructure wrecked, the *Gothic* turned back for New Zealand, Captain Agnew navigating by means of a chart and compass taken from a lifeboat. Late on 6 August the ship limped into Wellington Harbour under naval escort. It was patched up but scrapped the following year.

The *Gothic* limping back to Wellington show how the fire swept through the bridgework. Seven people died in the blaze.
Ref: EP-1968/3313/B5-F, Alexander Turnbull Library, Wellington, NZ

THE UNION COMPANY'S STEAMSHIP WAIRARAPA

8

Two of the worst — the *Wairarapa* and the *Penguin*

'As the party from the *Argyle* clambered over the rocks and waded round the jutting points to reach the spot, they found the bodies stretched out full length on the hard rocks. One of the women, both of whom had belts fastened around them, was entirely naked. The other wore a tattered chemise. The savage sharks had attacked the body of the former, which presented a shocking spectacle, the calf of one leg and a big piece out of her side having been bitten away. The younger girl was apparently 14 or 15 years of age, and looked a most pitiable object as she lay with upturned face on the sea-worn shingle, her eyes staring vacantly into space and her mouth filled with sand. The men in the party had to turn away their heads to hide their emotions as they took in the situation.'[1]

That article, printed in the *Otago Daily Times* on 5 November 1894, was about New Zealand's most widely reported nineteenth-century shipwreck. The colony was first horrified and then scandalised by the differences between survival rates for the crew (69 percent) and the passengers (45 percent). The repercussions were still flying thick and fast a month later when the court of inquiry attacked the master, his officers and crew. 'This conduct seems inhuman and inexplicable,' it thundered. 'Of Mr Fenwick, the purser, all we can say is that he saved himself.' The court was referring to New Zealand's third-worst shipwreck, the steamer *Wairarapa,* which ploughed into Great Barrier Island in the first minutes of 29 October 1894, killing 121 (some reports said 125) people.[2]

The *Wairarapa* was a steel screw passenger steamer of 1786 tons. Although no longer in the front ranks of the Union Company's fleet, the 12-year-old ship was still a mainstay of its trans-Tasman service. Like its sister *Manapouri,* the *Wairarapa* had been one of the world's first merchant ships to be fitted with incandescent lighting. The 13-knot vessel could carry about 200 passengers.

The *Wairarapa*'s career had not been without incident. In February 1884, while unofficially racing the *Adelaide* off Australia, it had collided with that ship, suffering about £31,000 worth of

OPPOSITE Alfred Ebsworth's woodcut of the *Wairarapa* showing the lavish interior.
The Australasian Sketcher, 4 April 1884, A/S09/04/84/60, State Library of Victoria

The light of day, and calmer seas, revealed the precarious position those on board the *Wairarapa* had found themselves.
Auckland Libraries Heritage Collections, 4-1023

damage. One *Adelaide* passenger, thinking his ship was sinking, leaped aboard the *Wairarapa*. Less than a year later the *Wairarapa* again made the news. Early on the morning of 1 November 1885 it caught fire between Napier and Gisborne. When the crew lost control of the fire, Captain H.W.H. Chatfield raced to Gisborne where the fire brigade took over. No one was killed, but the second stewardess burned herself while helping passengers. Fire-damaged along a third of its length, the *Wairarapa* was sent to Port Chalmers for £36,000 worth of repairs.

On 24 October 1894 the *Wairarapa* left Sydney on the start of what should have been another regular trans-Tasman voyage. Aboard were 65 crew and 95 saloon and 90 steerage passengers (mostly Australians). On the bridge was the regular master, one of the company's most senior, 55-year-old Captain John McIntosh. The ship, recently refitted at Port Chalmers, was in perfect order.

On 28 October the *Wairarapa* began picking its way down the Northland coast in thick fog. Company regulations required masters to reduce speed in such conditions and to sound their fog horn, but for some reason McIntosh — usually very prudent — ignored his officers' requests to slow down and refrained from using the fog horn — allegedly because many passengers had gone to bed early to arrive refreshed in Auckland the following morning. These were fatal errors, since at precisely eight minutes past midnight on the morning of 29 October the *Wairarapa* slammed into a 200-metre-high cliff about a kilometre east of Miners Head at the barren and forbidding tip of Great Barrier Island.

McIntosh compounded the problem by ordering the engines reversed — his rationale being that the further up the beach the *Wairarapa*'s bow was the easier it would have been for passengers and crew to get off safely. But with the ship's bow just metres from the cliff, the steepness of the bank made it impossible for anyone to get ashore that way.

The ship rested on a narrow rock shelf. Conditions were terrible and were about to get a lot worse. A heavy sea was washing the wreck, and the night — still thick with fog — was pitch black. Where were they? No one knew — the officers thought they were at the Hen and Chicken Islands. As passengers and crew huddled on the cold decks in the foggy darkness, they could hear waves breaking against still invisible cliffs. And they were already dying as waves plucked them from their perches. More went about 11 minutes later when the ship slipped off the rock ledge and slid further into the water. To complete the terror, 16 horses broke loose from their stalls on the deck, killing or stunning passengers as the terrified animals struggled vainly to escape.

Annie Howsea was travelling with her 70-year-old grandmother, Mrs Forbach, and a friend, W. Baker. When the alarm sounded, the girls got up and assisted Annie's grandmother to the deck where the three put on lifebelts, slipping them on just minutes before a huge wave washed them over the side into the boiling sea. There they clung together until Forbach slipped out of her belt and Baker disappeared. Annie counted

The bravery of the *Wairarapa*'s stewardesses, Annie MacQuaid, Charlotte McDonald, and Elizabeth Grindrod, was heavily praised by the newspapers. Although nineteenth century seafaring was a man's game, large passenger ships employed a few stewardesses to look after female passengers. David de Maus produced this composite image to honour the *Wairarapa*'s stewardesses.
Ref: 1/2-016394-G, Alexander Turnbull Library, Wellington, NZ

herself lucky, because she had been struggling to tighten her belt. For a long time, she clung to her grandmother, who implored her not to leave her and talked about them meeting death together bravely. Some time later that night the old woman died, later slipping from Annie's tired fingers. After what seemed like ages, the young woman was picked up by a boat and taken ashore.[3]

Individuals displayed great courage. The three stewardesses were everywhere, helping women and children into lifebelts and even giving their own belts to those without them. Dunedin jeweller A.J. Lumley surrendered his to stewardess Lizzie Grindrod before going over the side. Luck was with him. He was pulled into a boat by a fireman named Neill and taken to Whangapoua Bay where another boat — under third officer William Johnston — was waiting.[4] Elsewhere, quick-thinking crew saved several lives by cutting loose the life rafts and pushing them towards people in the water. Yet all but two of the boats were smashed before they could reach shore. Those two stood by until daylight, plucking swimmers from the water. Within hours the only people still alive aboard the wreck were clinging to the fore and main rigging.

The bridge, symbolic of McIntosh's shattered authority, did not last long. For a short while — probably no more than quarter of an hour — 40 or so people shuffled about on it, then it, too, collapsed and was washed away, taking everyone with it. A man reported that one great wave washed over them all, scattering them into the surf. After a 40-minute struggle he made it ashore, but most were not so lucky. McIntosh had remained on the bridge after the collision but was not present when it collapsed. Just minutes earlier he had climbed onto the port rail and plunged headfirst into the sea.[5]

'When daylight came a shocking scene presented itself to the gaze of the survivors,' the *New Zealand Times* reported. 'Straight ahead rose up a frowning perpendicular cliff, the ill-fated steamer on her port side, bows elevated, a perfect wreck, and quite close to the cliff, dead bodies mingled with the wreckage and deck cargo, while the sea for a considerable distance around was covered with flotsam of all descriptions — portions of deck-houses, cases of fruit, horse-stalls, etc.'[6]

Two very brave seamen (one of whom was second steward Kendall) swam ashore with a rope with which he then pulled ashore all the remaining survivors, except for two who lost hold of the rope and were carried away. Last off was steerage passenger John Austin. He had scrambled back aboard earlier when the third lifeboat had capsized, joining two Chinese men in the rigging. He quickly decided that it would be suicide to linger there and he persuaded his companions to leave with him. 'The Chinese did not savvy for a long time, but before they could venture into the water they stripped, and sent their clothes in front of them.'[7]

On the island, survivors — mostly scantily clad — huddled on the rocks, sustained by nothing more nourishing than a few oranges washed up alongside them. Conditions were still awful and the sharp rocks made it hard for barefooted survivors to help the diminishing band of swimmers being dashed against the jagged rocks.

It took time for the news to reach the rest of New Zealand. The *Manapouri* had spoken to the *Wairarapa* at 1800 hours on the 26th but nothing further had been heard of the ship. By 1 November, the Auckland papers were reporting that 'some anxiety is felt regarding her'.[8] The previous morning the *Rotomahana,* which had left Sydney 24 hours after the *Wairarapa,* had docked safely at Auckland. Where was the *Wairarapa*?

No one knew, and the Union Company sent the *Waihora* and *Wakatipu* to near North Cape and the Three Kings *(Waihora)* and to the south *(Wakatipu).* But no one was too pessimistic. Machinery and tail shafts were less reliable than

OPPOSITE Survivors and rescuers at their temporary camp.
F. Pulman, Auckland Libraries Heritage Collections, 589-137

they are today, and it would not have been the first (or last) time that a ship was left drifting around the Tasman for days or weeks.

Then came dramatic news. The *Wairarapa* had run ashore at Great Barrier Island, almost in sight of its destination, and most of the passengers and crew were feared dead. Local settlers and Maori reported bodies in the sea and all along the island's beaches. When the coastal steamer *Argyle* reached Great Barrier on Wednesday with the company's Auckland branch manager, Thomas Henderson, together with a party that included an undertaker, it called in at Fitzroy Harbour, where the postmaster said that the bodies of two women had already been buried and Catherine Bay Maori had interred a further dozen victims. But how many had died? Estimates varied widely. One settler had made the tally, of which 40–50 had already been accounted for. The later best estimate was 125 casualties; 40 bodies were never found.

There were some pathetic sights. On Tuesday rescuers retrieved the body of a woman, still clutching her dead baby, from Fitzroy Harbour. The *Argyle* rescued third officer Johnston from a small boat and then passed the steamer's bridge floating off Catherine Bay. There Henderson learned that settlers had recovered 19 bodies from the surf.

Collecting bodies was not for the faint-hearted. The first corpse of an unnamed man was found with a lifebelt under his arms; his only clothing was a portion of his nightshirt. Almost beside him was a dead horse. About 100 metres further on, the landing party found its second corpse, that of a young woman. And so on … As each body was found, planks, hammers and nails were produced and rough, makeshift coffins were knocked up and a careful record was taken of the appearance of the bodies, clothing, rings etc. Several bodies had been mangled on the rocks. Just around the corner, the men came upon bodies lying under bright red pohutukawa.

The unpleasant job of body recovery took days. Identifying them and contacting next of kin took longer. Passenger lists did not tally and confusion reigned for some time. Each day the newspapers reported new bodies, items found and, increasingly rarely, of people confirmed safe and sound. As late as 30 November the *Otago Daily Times* carried details of Oamaru survivor F. Hastie's wallet, of documents and letters belonging to Charles White of Sydney, and of a reported dispute over the ownership of 13 sovereigns found on the body of George Bird.

All this raw, emotionally charged evidence ensured that the court of enquiry was followed keenly. Judge H.W. Northcraft, assisted by Captains M.T. Clayton and Andrew, blamed Captain McIntosh for the disaster. The *Wairarapa* had overrun its distance on its south-east course; McIntosh had failed to allow for his speed, the current, or the set of the sea pushing him off course. 'I am of the opinion that the S.S. *Wairarapa* was lost through Captain McIntosh and the first and second officers not taking a correct point of departure at the Three Kings, and not allowing for a current, which by the first and second officers' evidence, they should have been aware was running to the south and south-east', Northcraft concluded damningly.[9]

Concluding that the 'so-called boat drill is a mere farce', Northcraft roasted the crew for failing to make the best use of the 11 minutes between initial impact and the heavy list to port. When the order came to launch the boats, it became a free-for-all because the men had not 'even had this farcical boat drill for over six months'.[10] McIntosh had suspended boat drill over winter, saying that his crew knew what to do.

Northcraft added cowardice to the charge of incompetence. The *Wairarapa*'s crew failed to meet 'what might have been expected of them, what we always expect of British seamen worthy of that name when there are women, children and passengers in peril'.[11] Quoting the evidence of several crewmen, he condemned McIntosh for telling officers and men to get into the rigging while passengers remained on the wave-tossed decks. In concluding this section of his report, Northcraft quoted carpenter Peter H. Thompson: 'It was thought that the ship was going down, and my impression was that it seemed to be everybody for himself.'[12]

Referring to the chief officer's conduct in letting the raft with 12 people float towards the Needles while he retained two serviceable boats at Whangapoua, the court said that 'no censure we can pass is severe enough'. His three-day delay in searching the coastline between the wreck and Whangapoua was described as 'inhuman and inexplicable', especially in view of the fact that many of the bodies later found floating in the water were wearing lifebelts. How many of them might have been saved if a steamer had appeared earlier?

Also slated were the second and fourth officers, and the purser, the latter of whom Northcraft said, 'all I can say is that he saved himself'. The only ones praised were the engineers, the stewards and the stewardesses, especially Charlotte McDonald and Annie Macquaid, each of whom was last seen giving her lifebelt to a female passenger. 'The conduct of these noble and self-sacrificing women is beyond human praise,' he concluded.[13]

The court's principal recommendations were for all boats to be uncovered and swung out in foggy weather and for the rafts to be launched at the first possible opportunity rather than left until all the boats had been launched. Costs were divided between the various parties.

The Union Company conducted its own inquiry, the results of which Managing Director James Mills presented to the board in a highly confidential report dated 10 January 1895. They make interesting reading. Although one could suspect Mills of bias — he never protected the incompetent. Furthermore, because the report was secret, it was not intended to have propaganda value and so should be taken at face value.

Mills agreed with the court's basic findings, blaming McIntosh for a display of 'recklessness' totally out of keeping with that officer's 'reputation of being a careful and prudent navigator'. Since it

Inside the entrance to the Dunedin Northern Cemetery is this monument to the stewardesses of the *Wairarapa*, erected by Union Company employees.
Gavin McLean Collection

was known that some passengers travelled in the *Wairarapa* because of McIntosh's known diligence, Mills attributed his aberrational behaviour to health problems, with McIntosh having recently suffered from repeated attacks of influenza. In September-October, while his ship was being refitted at Port Chalmers, he had gone to Lake Wakatipu for a rest. Only later did the company learn that McIntosh had 'consulted Dr Coughtrey about his health, informing him that at times that his nerves were shaken and that at times he did not feel confidence in his own judgement.' Clearly the man had had a breakdown.

Mills disagreed with the other findings, especially the assertion that the crew should have had time to swing out the boats after striking. With just eight (six seamen and two catering staff) on deck at the time (most of the rest were either down below or asleep), Mills believed that little more could have been done before the *Wairarapa*'s list made it impractical to launch the starboard-side boats (and the funnel's collapse destroyed one of the port-side boats).

Mills also rejected the finding that it had been everyone for themselves. While conceding that 'the chief officer merits to some extent the censure passed on him, and perhaps, also the second officer and purser' — concessions more significant than he allowed — Mills found that 'the ship's company as a whole behaved exceedingly well under very trying circumstances, and did not deserve the sweeping criticism passed upon them by the court, while in individual cases great heroism was displayed'.[14]

Mills reserved his greatest criticism for the chief, second officer and purser. The second officer he almost excused on the grounds of age. The chief officer, however, failed to provide a good example at the wreck itself and compounded his offence by failing to dispatch a Maori-owned boat to the mainland on Tuesday for help. Both officers were sacked. Purser Fenwick had been worried by McIntosh's behaviour and had slept fully clothed in the saloon. After the wreck he had done nothing to compile a list of survivors or help identify bodies. Mills' final remarks were interesting:

> In reviewing all the circumstances connected with this unfortunate disaster, it must be borne in mind that the catastrophe was of a sudden and overwhelming nature, the ship heeling over within a few minutes after striking, and her decks being swept by heavy seas on a rough night, the darkness of which was rendered even more intense by the prevailing fog ... correspondents in the newspapers have urged that, under certain conditions, the officers should have the right to set the master's authority on one side, and themselves take charge of the ship. This is a point to which I have given much consideration. I find it is the almost universal practice in our fleet for masters to consult with their chief officers at all times in cases of doubt. I hesitate, however, to recommend any regulation authorizing officers to interfere with the supreme authority of the master ... such a regulation would lead to divided authority, and in many cases would undoubtedly be a source of danger rather than the contrary. One of our new rules, however, will convey an instruction to officers, providing that in cases where a master is taking a course that is dangerous, or in defiance of the Company's regulations, they shall present the same to the master and record the fact of such record, to be subsequently forwarded in writing to the Marine Superintendent. I feel sure that this will afford sufficient check upon recklessness on the part of masters, while at the same time it will not in any way relieve them of their responsibility.[15]

Mills also rejected reports that McIntosh had been trying for a record passage. Mills had always forbidden masters to race their ships, and McIntosh had no incentive to try for a fast passage.

And what of the *Wairarapa*? Even as late as the 1930s it would continue to fatten its file in the Union Company building as relatives tried to track down information about former passengers or crew. Rumours about the ship's safe being loaded into a boat and then lost in the surf drew the curious. In November 1923 W.E. Vear and J. McKinnon, working from the scow *Katie S* against the wishes of the Union Company, bought the *Wairarapa* from the Minister of Marine and started diving on it and the wreck of the nearby American schooner *Cecilia Sudden*. Using dynamite, they brought up about 40 tonnes of materials, including the propellers, some shafting and a six-tonne condenser. They found several horse skeletons on the seabed but no human ones and no sign of the safe. A boiler was discovered some distance away, full of crayfish.

In 1979 divers and authors Steve Locker-Lampson and Ian Francis reported that the ship was badly broken up but was still a wreck of importance. The starboard side is still semi-intact and large pieces of wreckage abound. Indeed, the compass had been recovered a decade earlier, still in working order.[16] In the twenty-first century, the wreck, while well picked over, remains popular with recreational divers.[17]

of 13 February 1909 reported. Anxious relatives or friends of passengers and crew members clustered around the notice boards and swept from room to room. Some — when the wires and messengers failed to bring news about the safety of a father, mother or child — hoped on for better tidings, but the hope of others broke down, and the sounds of women's sobbing rose above the murmur of voices asking, answering, wondering about the chances.'[18]

The paper was talking about one of the costliest shipwrecks in our short history. The ship was the *Penguin* — the ferry that ran between the capital

The wreck of the *Penguin*

All day the Union Company's Wellington office was crowded with people 'eager for the latest word about the missing and the saved', the *Evening Post*

By the time the *Penguin* was running across Cook Strait on the ferry service, as illustrated here, the original ship had been rebuilt and fitted with more accommodation space.
Gift of Wellington Museums Trust, New Zealand Maritime Museum (2012.0.1419)

and Picton and Nelson. Although initial fears that all but 13 of the 105 aboard had drowned were false, the loss of 72 lives was a disaster.[19]

At first the *Penguin* seemed an unlikely victim. True, it was rather long in the tooth, having been built at Glasgow in 1864. Since coming to New Zealand in 1879, however, the *Penguin* had been well maintained. The little 824-ton ship had plied many of the company's most important routes. In 1892 it was refitted with more powerful and economical engines. Three years later the *Penguin* commenced weekly return voyages between Wellington and Lyttelton, inaugurating the famous 'steamer express service'. In 1902, after a brief stint in Australia, the *Penguin* returned to the passenger/cargo service between Wellington and Nelson and Picton.

The steamer's last voyage began quietly. At 1820 hours on 12 February 1909 the ship left Picton for Wellington under the command of Captain Francis Naylor. The 36-year old former Londoner had been with the Union Company since 1897 and was widely respected as a reliable and careful ship handler.

Two hours later, when the *Penguin* left Tory Channel and entered Cook Strait proper, the weather was clear, though threatening to deteriorate. The strait, one of the roughest, windiest stretches of water on earth, is notoriously fickle. In minutes it can change from calm and sunny to a howling gale — as it did that day. By the time the ship was halfway across, the weather was too thick for there to be any chance of picking up

Minor strandings were common for ships in the colonial and Edwardian eras. The *Penguin* experienced many such incidents around Picton, the Sounds and Nelson. 'A large number of people visited the scene during the day, and the cabs and buses have been doing a thriving trade', the *Nelson Evening Mail* reported of one such minor stranding on 26 November 1895. 'Local photographers were also very much in evidence.' The light at Haulashore Island had been out at the time. The ship was refloated undamaged that same afternoon and continued on to Wellington.
Ref: PA7-01-54, Alexander Turnbull Library, Wellington, NZ

Clearing the beach of wreckage and bodies at Cape Terawhiti, Wellington, in the days after the *Penguin* was wrecked. Note the upturned life-rafts beneath the cliffs — many of which had suffered major damage.
Ref: 1/1-020152-G, Alexander Turnbull Library, Wellington, NZ

the Pencarrow light. At 2140, therefore, Captain Naylor set an outside course, intending to take the *Penguin* well clear of Sinclair Head and Thom's Rock. He kept up speed as the sighting of another ship on his starboard bow reassured him that he was clear of danger (he had allowed for a strong southerly set).

At 2202 Naylor changed his mind and ordered a change of course to take the *Penguin* right out to sea. But it was too late. The *Penguin* struck a rock with a grinding crash, sliding along the submerged object on its starboard side. Seaman Charles Jackson thought it sounded just like 'the rending of a gigantic piece of calico'.

Although the court of inquiry would later say that it was 'far from satisfied that the vessel struck on Tom's Rock [sic]' and that it could not locate with certainty the scene of the wreck, it was widely believed that Naylor had hit Thom's Rock (Toka-haere), approximately abreast the outfall of the Karori Stream. Thom's Rock is one of many hazards on this rocky coast. Usually submerged by about 1.5 metres, the rock is the outermost of the Terawhiti Rocks, and the most exposed. A vicious current of 3–6 knots sweeps past and anyone who survives the initial impact has to contend with this as well as the currents that develop closer to the rocky shore. Several ships, including the barque *Grasmere* in 1895, had come to grief in the vicinity of Thom's Rock.

But back to 1909. Naylor swung out the boats as soon as soundings revealed that the chain locker and the forecabin were flooding. Soon the Nos 1 and 2 holds were filling and it was obvious the ship was doomed. Beaching was out of the question — the coastline was too exposed and rocky. Once the initial shock passed, officers and men started putting women and children in the boats.

The first boat dropped bow-first into a heavy sea that swamped it, flinging the hapless occupants into the heaving waves. The people still aboard the *Penguin* gasped with horror, but most of the floundering passengers and crew made it back onto the deck. Soaked and scared, they reboarded the lifeboat with instructions to pull far out to let the flood tide take them around Terawhiti, but two boats sank before they could get very far. Naylor was soon too caught up in his own personal struggle for survival to be of much help to others. Sensing that his shattered ship would soon sink, he stepped into the sea from his bridge, ready to accept his fate. A wave drew him towards wreckage from one of the lifeboats that supported him as he drifted along with the current and enabled him to observe how others were faring in their struggle for survival.

'I could see the rafts. They had a very rough time,' Naylor recalled, while fighting to retain his hand-hold. 'They got into a swirling, treacherous sea, and were swung round and round. Several times those who were on them when they left the ship were washed off,' he remembered. 'I believe all on the rafts reached the shore,' he told a reporter.

The sea dumped Naylor on the beach about three in the morning. Two rafts came ashore near him, with 11 and 12 survivors respectively. All had had a very trying time, one raft-load being overturned completely three times. It was a bleak landfall. The night was pitch dark, the rain was heavy and the southerly was shrieking against the cliffs. Since it was too dark to risk going any further, Naylor crawled into the scrub and curled up, waiting for dawn. Other luckier survivors from the raft made it to the station of a settler named McMenamin, who soon had fires blazing and food and drink on the table.

While the rafts had been heading towards the shore, the occupants of the lifeboats had been fighting their own battles. For some it was a brief struggle. As we saw, the first boat launched was smashed to pieces. The second fared little better. Partly stove-in, it nevertheless looked as if it was going to get away successfully. But just when it looked as though the boat had made it, a heavy sea capsized it, drowning most of those aboard. The other boat, launched with great difficulty by Jackson and the second officer, also turned turtle within minutes. The crew tried to launch the No. 3 lifeboat but by then the ship's list was too steep to permit it.

It was one of the night's saddest ironies that the lifeboats, which probably looked safest to the passengers, gave the poorest chance of survival. Only two people reached shore alive on one, and they did it under the craft, rather than in it. When rescuers dragged a broken lifeboat out of the surf, they rolled it over to discover a youth named Ellis Matthews, and Ada Hannam — a passenger from Blenheim. Hannam, who lost her husband and four children in the disaster, was the only woman to survive.

When disaster struck, Ada Hannam and the other women were called to board the lifeboat. Stewards handed out blankets, but Hannam was not flustered by the atrocious weather or the ship's state. Seconds later, as the crew struggled to lower the boat, the tackle collapsed, dropping it into the water, bow-first. Three of her children — Ralph (10), George (five) and Amelia (three) — were swept away in seconds. They cried out to her but there was nothing she could do for them.

For the rest of the night the boat drifted towards the shore. Once she sighted a raft and called out but received no response. Then the rocks loomed up. The only seaman aboard tried to keep clear of them with his oar, but a big wave struck the boat, capsizing it and throwing everyone into the water.

'When the boat turned over, I clung to it again. I could see my baby was still tied to the seat. There was a young boy named Matthews. I kept hold of him, for we were then underneath the boat,

Hannam recalled. 'We both clung to the seats but could see nothing, for the boat covered us and our heads touched the bottom of it. The air in the boat kept us alive, because every time there was a big wave, it lifted the boat a little and let in the fresh air.'

She calmly accepted the news that her baby, Ruby, whom she had tied to the boat, was dead. 'Oh, let me untie it myself,' she told the men nervously offering assistance by untying the tiny corpse, 'I know how I fixed it up.'

So much for the lifeboats — what about the rafts? Once the boats had gone, the crew unlashed the rafts and pushed them over the side, shepherding the remaining male passengers aboard. By now the *Penguin*'s engines had stopped, and, its bows well down, the ship was wallowing as each wave hit. Jackson, being a strong and confident swimmer, was the first to leap towards a raft. 'I stuck to the painter, and I looked up at the vessel. She was down by the nose with her stern high out of the water,' he recalled. 'Some of the men on the deck were afraid to jump but they soon recognised that the ship was fast settling down. One by one they commenced to leap over the side, and happily, all managed to reach the little craft that was being buffeted about in the waves. Altogether 12 people clambered aboard the raft, of which I took charge.'

They only just made it. Moments later the *Penguin*'s boilers burst. Cold water rushed in to meet red-hot iron, causing a tremendous explosion that tore the ship apart. Fortunately, the vicious seas had swept both rafts out of the danger zone from falling debris or from being sucked under.

Survivors from the first life raft. Front, left to right: able seaman George Farrell, chief cook D. Lynn, second engineer William Luke, purser Arthur Thompson and steward David McCormick; rear, passengers A.L. Hopkins, Gerald Bridge and George Perkins.
Gift of Wellington Museums Trust, New Zealand Maritime Museum (2012.0.9079)

Jackson's passengers found two oars in the water and used them to guide the drift of their pitching and tossing craft. For three hours they drifted in the dark, overturning three times. Fortunately, they were all strong swimmers, so they scrambled back aboard each time. Their ordeal ended around 0300 hours when they glimpsed the dim outline of the shore through the haze and mist. Then, with a final rush, the raft thudded on to the rocks, enabling them to scramble ashore, cut and bruised, but glad to be safe.

Many people had not made it that far. Jackson's last glimpse was of Captain Naylor, who was floating nearby, on the battered fragment of a boat. 'He was accompanied by a passenger, apparently a young man, between 25 and 30. By some means, of which I am not aware, this poor fellow had one of his hands torn from the arm, and I myself saw the Captain taking out his handkerchief and binding it above the poor fellow's wrist,' Jackson recalled. 'The captain managed to reach the shore, as we did, but his companion did not. His dead body is now lying on the beach.'

There were enough of them piling up. For the next few days police and volunteers scoured the coastline looking for survivors or bodies, finding none of the former but plenty of the latter. In all they recovered 61 bodies, fewer than half of which were identified. Today they lie in the Karori Cemetery, where visitors can follow a self-guided *Penguin* walking tour booklet.[20] In 1909 sightseers had a field day. Mud and the swift rain-swollen streams made it a daunting expedition, but many Wellingtonians nevertheless, made the rough journey to gawp and gaze. One of these sightseers was 'Zak', Joseph Zacharia, a well-known Wellington photographer whom the press considered ghoulish for taking pictures of the dead. But with so little wreckage visible, what else could he record?

Perhaps inevitably, the court of inquiry slated Captain Naylor for his actions leading up to the disaster. Although it conceded that

On 12 February 2009 — the centenary of the disaster — the Wellington City Council unveiled a plaque commemorating the wrecking of the *Penguin*. The plaque is located on a prominent rock at Tongue Point, close to where the ship is thought to have hit rocks and foundered. While off the beaten track, the site is accessible by foot and four-wheel-drive.
Wellington City Council

the presence of an exceptionally strong flood tide had contributed to the tragedy, it criticised him for not putting his vessel's head to sea soon enough after he had run a course of 18 miles. On a majority verdict (one of the nautical assessors dissenting from some aspects of the report), the court suspended his certificate for 12 months. He appealed, raising the possibility that the ship might have hit the wreckage of the brigantine *Rio Loge*, lost somewhere off North Canterbury a month earlier. The ship had been carrying timber, which some believed may have kept the wreck semi-submerged until the *Penguin* hit it.[21] Indeed, some observers reported finding a surprising amount of timber scattered along the shore. Although the court stated that it was not necessarily convinced the *Penguin* had hit Thom's Rock, the appeal court dismissed Captain Naylor's appeal and the likelihood is that the first court got it right.

PART THREE
Foul Play

9

Fraud, vandalism and terrorism

As insurance companies know, deliberately wrecking ships in order to claim insurance money is an ancient practice. British maritime history is full of 'coffin ships' — overloaded old migrant ships sent to sea in the hope they sank — as well as cut-and-dried cases of deliberate wrecking.[1]

In the nineteenth century the sea claimed so many ships that less scrupulous shipowners and speculators must have been tempted to abandon ships on the other side of the globe, away from effective detection by insurers. People still try it on. In the 1980s Lloyd's of London was plagued by a succession of tankers sinking or being abandoned off the African coast.

The very nature of the crime made documented cases of premeditated wrecking rare. Suspicion abounds in several New Zealand cases — one as recently as the 1960s — but hard evidence is lacking. The earliest deliberate wrecking may have been the brig *Wanderer* at Riverton in early 1861, but evidence is sketchy. Ingram quotes an article on early Riverton which hints discreetly that: 'ships had a habit of coming to stay as when the brig *Wanderer* gently eased up on the beach and settled to rest. As the captain pointed out, she had had 24 years of service and was entitled to a pensioned old age.'[2]

Robert Fulton was less cautious in *Medical Practice in Otago and Southland in the Early Days*. Writing about a local general practitioner, Fulton recorded that:

Dr Monckton found that vessels came to New Zealand for the express purpose of stopping there, and on one occasion the brig *Wanderer* with a load of grain, came ashore in a perfectly calm sea only two miles from Riverton. When the tide receded, the men stepped over the bows on to the beach and

OPPOSITE The Union Company's *Hawea* had many mishaps during its short but eventful career. One such incident took place at Nelson on 30 March 1886 when 'the residents near the harbour entrance awoke this morning and drew their blinds (and) they were surprised to see a large steamer with her nose on the beach at the beacons opposite Mr Richardson's house.' At 0415 the *Hawea* had failed to answer the helm and ran up on the shingle under 'easy' steam. The shingle was described as soft and the ship was undamaged, so after disembarking the passengers and redistributing the 25 tonnes of cargo from the forehold to the stern, the crew waited for the next high tide for the steamer *Mahinapua* to pull the ship from its embarrassing 'berth'.
Ref: 1/4-002679-F, Alexander Turnbull Library, Wellington, NZ

PAGES 124–125 The image that flashed around the world: the *Rainbow Warrior* listing against Marsden Wharf, the banner above its wheelhouse still defiantly demanding a nuclear-free Pacific.
New Zealand Herald

camped under tents made of sails until the master had arranged for unloading and carting the cargo away. Dr Monckton had a look at her and being a bit of 'a practical man', and always inclined to mix with business outside of his own, offered to get her off the beach and into the river for £25. Captain Howell, to whom he made the offer, laughed at him, saying that he knew it was quite practicable, and would cost half that; but as there was no Lloyds Agency in the neighbourhood to make objections on behalf of the underwriters, the wiser course was to let the brig rest her poor old bones where they were. She had been floating for 27 years, and the doctor would be a wise man not to interfere in other people's business.[3]

Our next example concerned the loss of the Union Steam Ship Company coaster *Hawea* at New Plymouth in 1888. The *Hawea* was one of the new steamers with which the company started business in 1875. At 721 tons and powered by new, efficient, compound engines, the *Hawea* and her sister ship *Taupo* set new standards. When Union's managing-director, James Mills, and chairman, George McLean, puffed up the coast between Dunedin and Onehunga, they were feted at every port. 'Travellers, unless their business is very urgent, elect to wait a day or two in order to secure passage by them,' one paper observed.[4]

Bigger ships followed, but the *Hawea* was still a valuable asset 13 years later when, late at night on 11 June 1888, it arrived off New Plymouth from Onehunga. New Plymouth, like so many minor ports, was going through a bad patch. The harbour board was divided and the construction of the breakwater was behind schedule. Seafarers treated New Plymouth with caution, so Captain John Hansby anchored the *Hawea* to wait for dawn. At 0630 hours he began to approach the wharf. A strong northwester was bouncing the ship about, and the anchor slipped off its chain while being raised. Undaunted, Hansby continued towards the breakwater wharf before reversing briefly to make room for the coaster *Gairloch*.

While reversing, the *Hawea* touched 'on her heel slightly aft', then struck again more heavily.[5] Hansby stopped the engines, but the *Hawea* kept striking heavily aft and started flooding, settling by the stern with its bow poking high out of the water. Hansby gave the order to abandon ship and the 11 passengers, their baggage, and the mails were put in a boat and rowed to the wharf. When loading the second boat, the *Hawea* lurched almost on to its beam ends, sweeping away chief officer William Waller. A whaleboat rescued him, but Allegro, a prize racehorse being unloaded, drowned.

By 0930 it was obvious the *Hawea* was doomed. Water pressure burst the engine room bulkhead and the steamer rested on an even keel. After seeing off the remaining hands, Hansby and Waller left the ship. The Union Company's local agent, W. Newman, supervised the removal of the compasses and minor deck fittings before telegraphing Dunedin: 'Weather now deteriorating, shall pay off all hands tomorrow.'

A napkin ring salvaged from the wreck of the *Hawea* in 1930, just before the wreck was intentionally demolished by explosion. The location of the wreck meant that it had been a headache for dredges, who over the years had unwittingly brought sections of it to the surface during work to deepen the waters around Newton King Wharf, some 100 feet from where the *Hawea* had gone down.
PA2004.195, Puke Ariki

The storm dashed hopes of salvaging the wreck; eight days later it fetched £120 at auction and the fittings around £90.

Although the *Hawea* was insured for £13,000, and the court of inquiry (which could not discover the cause of the disaster) exonerated Hansby, the Union Company sacked him. The indignant captain blasted his former employer publicly, alleging the accident arose...

> solely through the fault of the 'so-called' harbour, or instructions of their servants, and did I think them worth the powder and shot, might possibly resort to other means to vindicate myself. However, I will let a 'sleeping dog lie' and do not intend by controversy to bring grist to the mill of the *Taranaki Herald*. Let them take any action they like to prove that I struck anything harder than water outside their 'so-called' harbour and I will be quite prepared to fight the battle again.[6]

Hansby defected to the opposition and skippered the trans-Tasman trader *Jubilee*. An unlucky man, he was run down by a tram in Sydney a year later.

In fact, despite all the finger-pointing, no one wanted to make an issue of the accident. The Union Company threatened to claim £30,000, alleging that the harbour board's staff acted negligently in signalling the *Hawea* to enter port and by failing to clear the channel of obstructions, but ultimately took no action. It knew that it would have been too risky to push things.

Why? The harbourmaster insisted that there was five metres of clear water where the *Hawea* was swinging, and Newman was not inclined to dispute this. At the preliminary inquiry on 13 June, the harbourmaster had mentioned the possibility of the *Hawea* overriding its anchor, but at the full court of inquiry a week later (while still employed by the Union Company) Hansby had stressed that it would have been impossible for the ship to override its anchor as wind, tide and helm

Despite all the finger-pointing, no one wanted to make an issue of the accident. The Union Company threatened to claim £30,000, alleging that the harbour board's staff acted negligently in signalling the *Hawea* to enter port and by failing to clear the channel of obstructions, but ultimately took no action.

were against that.[7] But were they? As Newman advised James Mills confidentially: 'I think it would be unwise to employ a diver to look for the supposed obstruction ... as circumstances would indicate that it could not have been anything but the anchor which did the damage and so possibly open up the question of damages against ourselves.' Although a rock had been found 76 metres astern of the wreck, Newman did 'not for a moment think that it contributed to the incident. ... It would be policy no doubt to give the Harbour Board formal notice of action, which will obtain publicity and may prevent Keith [owner of the valuable racehorse Allegro] from pursuing his suit against us for the loss of Allegro.'[8]

And so the company let its claim against the harbour board lapse. The insurance companies paid up. Although the board several times debated removing the wreck, and F. Oldfield pottered about in 1890–91, raising more hot air than wreckage, the *Hawea* sank into the sand and into obscurity. The last evidence was blasted apart during interwar dredging operations. The *Hawea*'s bell is in the city museum.

The *Ariadne* scandal

New Zealand's most infamous case of deliberately casting away a ship occurred at the mouth of the Waitaki River on the evening of 24 March 1901. The vessel was the 230-ton schooner-yacht *Ariadne*. Built in a lavish manner at Gosport, England, in 1874 of oak, teak and pine, the *Ariadne* was the nineteenth-century equivalent of a modern super yacht. Fast, stylish and elegant, it flew the flag of the Royal Thames Yacht Club and, when built, was described by its owner as the largest schooner-yacht sailing under the British flag.[9] Luxuriously appointed, the *Ariadne* had won the German Emperor's Cup at Cowes.

But after schooner-rigged yachts went out of fashion, the *Ariadne* was sold to Thomas Kerry for £2100, a price he described as 'more a gift than a purchase'.[10] The dawn of the new century found the ship in the Pacific, on a voyage already marked by trouble. The *Ariadne* had grounded on a reef off Thursday Island and had been refloated with difficulty. When it limped into Sydney in November 1900, Kerry paid off the master and the crew. Shortly afterwards he hired a new master, George Mumford, who had been recommended by former crewman Andrew Olsen, and instructed him to sail the *Ariadne* across the Tasman to Dunedin to reprovision. Business prevented Kerry from sailing aboard on 25 February.

The *Ariadne* had been beating down the coast on the final leg of an uneventful trans-Tasman voyage when at 2000 hours on the evening of 24 March the helmsman saw breakers. Within minutes the *Ariadne* hit a shingle beach, swung round smoothly, and lay broadside to the shore. Although they did not know it at the time, they had gone ashore about three kilometres south of the Waitaki River mouth.

It was a gentle wreck. The *Ariadne* hit at near high water, running so far up the beach that the crew stepped ashore in knee-high water. At first it was hoped that this might save the ship, which was intact. The weather improved in the following days and Port Chalmers businessman John Mill, who bought the wreck for £215, thought he had got a bargain. So did the Kai Tahu from the Waitaki reserve who were charging sightseers a toll to gawp at the schooner stuck in the shingle. It was too good to last. Heavy seas over the next few days reduced the *Ariadne* to matchwood, scattering its expensive debris for kilometres. Mill went away happy, though: he had salvaged the masts and enough lead ballast to make a profit.

The initial inquiry — held at Oamaru under the jurisdiction of Major Keddell SM — heard that Mumford's charts were inadequate, the crew was unfamiliar with the coast and that the *Ariadne* was 48 kilometres off course when the accident occurred. Keddell found Mumford negligent of his responsibilities to his employer's interests and guilty of a grave error of judgement in not wearing (tacking) his ship two or three kilometres sooner than he did. He suspended his certificate for three months and fined him 15 guineas. Although an April hearing at Dunedin explored whether the ship had been 'designedly stranded', it found no evidence of fraud.

Case closed? Not quite. Waterfront rumours had it that the ship had been wrecked deliberately. Kerry had not helped things by mentioning high values for the ship, at different times reckoning its value at £15,000 or £20,000, nearly ten times what he had paid. Captain Stewart Willis — Lloyd's agent at Christchurch, who estimated the *Ariadne*'s value at £5000, twice the sum it was insured for — grew suspicious. Things did not seem right. Was not the loss of the only chart Mumford was using just too convenient? And what about those reports of the lockers normally used to store valuable items such as plate, linen and furniture being empty?

The elegant schooner-yacht *Ariadne* ashore near the mouth of the Waitaki River, North Otago, in 1901.
Ref: 1/2-123650-F, Alexander Turnbull Library, Wellington, NZ

Lewis hired Detective Fitzgerald, who advised him to come to Dunedin, where Mumford was shooting off his mouth. Lewis met Mumford at Dunedin's opulent Grand Hotel, where the agitated mariner spilled the beans: at Sydney, Kerry had asked him to wreck the ship in New Zealand before the insurance policy expired. They had planned to do this off the West Coast but had changed their minds after a storm washed away the lifeboat. Mumford said that Kerry had offered him £400 for wrecking the ship and had promised him the command of a larger ship he hoped to buy in England with the insurance money. This, too, he hoped to wreck later. Mumford offered to put this confession in writing for £400.

Why would he do something so risky? Mumford professed to be angry with Kerry, now safely in Sydney, for not paying him the first instalment of his fee and seemed to want to get even, whatever the cost. In a bizarre letter to his former employer, Mumford repudiated their agreement (he claimed to have lost his copy of a written agreement he had received from Kerry) and threatened retaliation through Willis and the authorities. A few days later Mumford moved to Christchurch under the alias 'Captain G. Stevens', worried that Kerry might pursue him.

On 7 June Mumford presented Willis with the wrecking agreement which, he claimed, had fortuitously turned up in the lining of an old coat. Purportedly signed by Kerry, it read 'I, T. C. Kerry agree with George Mumford to pay him £12 wages as master of the *Ariadne,* and a further £400 if the vessel is totally wrecked.' Quite an unusual set of articles!

Although Willis found this document difficult to believe (Mumford's letter to Kerry had indicated that there had been no agreement), an investigation showed that the signature appeared to be Kerry's. Willis, who until then had merely been collecting evidence with a civil case for his principals, Lloyds, paid Mumford and then went to the police who arrested Mumford on 8 October at Lyttelton. That same day Sydney police arrested Kerry and Eric Freke — Kerry's 18-year-old companion also implicated by Mumford. All were charged and bailed on the rare charge of deliberately casting away a vessel.

The affair now took on the characteristics of a bad penny dreadful. After Sydney barmaid Annie Downing said that she had overheard the men plotting to wreck the ship, Mumford met her at a Christchurch hotel and attempted to get her to retract her testimony. Alleging that he had spoken of wrecking the *Ariadne* while drunk, the increasingly irrational Mumford promised that Kerry would pay her well if she did as she was told.

But he was, in the language of Edwardian melodrama, thoroughly undone. Two detectives listening to this highly incriminating conversation from behind a curtain dragged Mumford off to charge him with attempting to bribe a witness. This time there was no bail.

The trial opened before Mr Justice Denniston at Christchurch on 20 January 1902. Kerry had hired the cream of the colony's legal brains to defend him, A.C. Hanlon and Charles Skerrett.[11] All the accused pleaded not guilty. The Crown outlined the case against the men, adding that Kerry had increased the suspicion of his involvement in unlawful activities by removing about two tonnes of valuable items from the ship, including binnacle lamps, a sextant and a chronometer case just before the *Ariadne* sailed from Sydney.

The third day of the trial was the turning point. That day the Crown produced scientific evidence to show that the document produced by Mumford had been traced over (in other words, it was a duplicate of one found in Kerry's possession) and that the incriminating words supposedly ordering Mumford to wreck the *Ariadne* had been added. Further damning evidence came from a waiter in a Christchurch hotel who said that he had seen Mumford writing intensely on a paper marked 'agreement'. At that point charges against Freke were dropped. Kerry also benefited from the exposure of this forgery and by the ploys that Willis had used to obtain evidence. When Kerry's counsel, Alfred Hanlon, asked why Willis had not suspected the agreement to be a forgery, since Mumford had first said that Kerry had not given him a written agreement, the case against the shipowner crumbled.

The jury found Kerry innocent but the now discredited Mumford guilty. Denniston sentenced him to four years' hard labour. Kerry returned to England with Freke, who was probably his lover. The case, a sensation at the time, left many doubts, and the *Ariadne* story continues to feature in books and television programmes.

Modern villainy

Since the end of World War II ships have been lost through violence twice (three times if a 10-metre fishing boat blown up at Opotiki in a drugs feud in November 1983 is included). The first incident was really just a bit of bungled petty theft. The second sent shockwaves around the world.

On 2 June 1966 the explosion that sank the derelict 272-ton steam trawler *Hautapu* at its moorings in Shelly Bay, Wellington, ended its undistinguished career. Launched at Port Chalmers in 1943 as a steel minesweeping trawler, the *Hautapu* became a commercial trawler after the war. That came to an end on 2 November 1963 when the *Hautapu* struck an object off the Kaikoura coast and was beached in Clifford Bay, south of Cape Campbell where the crew got ashore in rough surf. There the ship lay, battered by the sea and trashed by vandals, until April 1964, when it was refloated and towed to Wellington where it was found to be uneconomic to repair. Shunted around to Shelly Bay, the *Hautapu* gathered rust, blighting genteel harbour views.

The air force planned to use it as a bombing target but was thwarted when HMNZS *Inverell*, meant to tow it to the firing range, was diverted to search for survivors of the *Kaitawa*.[12] Then 'a couple of water front rogues', planted a bomb in the engine room and the hulk sank stern first, its bow still afloat. There it lay while NZ Fisheries Ltd — which argued that its responsibilities had ended when it gave the ship for use as a target — defence officials, and the harbour board, squabbled over who would pay to remove the wreck. Eventually the need to transfer the Shelly Bay complex to the harbour board forced a compromise: navy divers, assisted by the floating crane *Hikitia*, started cutting it up in 1972. Within a year the *Hautapu* was just a memory.[13]

The last, and undoubtedly the most sensational, attack in recent years was that of the French secret service on the Greenpeace ship *Rainbow Warrior* at Auckland on 10 July 1985. Shortly before midnight a loud explosion rocked the peace of Auckland Harbour. A second blast followed two minutes later. It was all over in minutes.

The *Rainbow Warrior* had started life humbly in 1955 as the British research trawler *Sir William Hardy*. For over 20 years it sailed in and out of Aberdeen, catching fish and conducting experiments for the Torry Research Station. By 1977, however, the scientists needed a bigger ship, and it seemed that the old trawler's days were numbered. With Britain's fisheries in decline the remaining companies had switched over to stern trawlers.

Then the vessel was discovered by enthusiasts from Greenpeace. Canadian David McTaggart purchased it in 1977, refitted it, renamed it *Rainbow Warrior* (an Anglicisation of the French name *Le Combattant de l'Arcen-ciel*, not a reference to the colourful Cree legend so often quoted) and sent the ship to sea to campaign against nuclear waste dumping and whaling. Greenpeace worked the ship hard until 1983–84 when engine problems and old age made themselves apparent. By now Greenpeace was acquiring bigger, better ships for its growing eco-navy, and the *Warrior*'s

The ship no one wanted. The old steam trawler *Hautapu* rests on the bottom after a botched attempt to steal some fittings. In July 2007 the *Dominion Post* reported that Wellington woman Shirley Thomas revealed that her brother-in-law Roydon Thomas was one of two men who accidentally sank the old hulk. One night, while drinking, they decided 'that it was a crying shame that all the brass was going to a waste'. They sneaked aboard and laid a few small charges to blow some brass portholes off their mounts. Unfortunately, these small blasts set off the larger one that had been packed on the ship for its official scuttling in Cook Strait. 'Roy said they were standing on the wharf nearly crying that it was going down', Shirley Thomas recalled.
Evening Post

defenders had to argue hard to have it sent to Jacksonville, Florida, to be re-engined, fitted with tall masts and auxiliary sails and smartened up.

The ship's new role was to draw attention to the plight of indigenous people affected by American and French nuclear testing in the Pacific. It had probably done its work too well for its own good by the time it motored up the Waitemata Harbour in July 1985. The French government, facing mounting criticism for its colonial policies and for testing nuclear weapons in the Pacific, had had enough of New Zealand-based anti-nuclear protests. It did not relish the prospect of the *Rainbow Warrior* leading another protest flotilla to Moruroa, the Pacific atoll that it used to test its weapons. Unknown to anyone, a small group from the DGSE, probably supported by the French navy, began planning to sink the vessel. In an act of unprecedented state-sponsored terrorism, teams of French agents infiltrated New Zealand, hired scuba gear and planted mines against the hull of the old ship while it lay alongside Marsden Wharf.

At about ten minutes before midnight the first explosion rocked the ship. Crew members and supporters, many asleep at the time, scrambled ashore, shocked and dazed. All escaped except Portuguese photographer Fernando Pereira who decided to go back to retrieve his camera gear. The second explosion, about two minutes after the first, trapped him in his cabin and finished off the *Rainbow Warrior,* which sank by the stern. Pereira drowned; his body was recovered by divers four hours later. Another crew member, Hanne Sorensen, at first feared missing, turned up from a walk in the city to see the *Rainbow Warrior* listing at an angle of 30 degrees to starboard, its stern resting in the Waitemata's mud.

Initial thoughts that the explosions might have been accidental — empty fuel drums or oxyacetylene equipment — were dispelled when divers saw that the hull plating had been bent inwards by the force of the blast. This was clearly no internal explosion. There was a hole about two metres by three metres on the starboard side of the ship's engine room and a second one near the propeller shaft. Someone had used explosives. But who?

Although some blamed the CIA, and others the opponents of French anti-colonial movements, the French government was the prime candidate. It was discovered that France had infiltrated Greenpeace while other DGSE operatives had reconnoitred Auckland. In the days and weeks that followed, the police investigation, the biggest in New Zealand's history, turned up tales of strangers with foreign accents, the hired yacht *Ouvea*, Zodiac dinghies, diving gear and hired camper vans. Then the DGSE agents Major Alain Mafart and Captain Dominique Prieur, masquerading as Swiss tourists, Alain and Sophie Turenge, were arrested trying to flee the country. On 24 July they were charged with passport fraud, murder, arson and conspiracy to commit arson.[14]

The long court cases, international lobbying and power politics are outside the scope of this book; suffice it to say that the French government eventually had to compensate Greenpeace and the New Zealand government.

The *Rainbow Warrior* was the most politically significant maritime disaster in New Zealand's history. Until its sinking, the Labour government, elected in 1984, had been struggling to reconcile two politically diverging policy aims: satisfying the electorate's firm anti-nuclear feelings, while remaining part of the ANZUS alliance. As Prime Minister David Lange recalled, France's criminal behaviour drew no criticism from New Zealand's traditional superpower allies, Britain (led by Margaret Thatcher) and the United States of America (led by Ronald Reagan). 'Eager as they were to condemn terrorism in its other forms, they were curiously silent about events in Auckland,' Lange recalled in 1990. 'The leaders of the West expressed not a moment's outrage about terrorism directed by a government against opponents of nuclear deterrence.'[15] With friends like them, who needed enemies?

The navy refloated the *Warrior* which, now fit only for scrap, lay alongside the Western Viaduct for the next two and a half years. Salvage was not easy. 'Excluding the bomb damage, which was extensive, the vessel was thirty years old and poorly maintained,' a Ministry of Transport official involved recalled. 'Many unofficial modifications had been made.'[16] But the navy, assisted by other

You can't sink a rainbow? The stripped hulk makes its way under tow to its scuttling ground off the Cavalli Islands in December 1987. *New Zealand Herald*

agencies, had the *Warrior* afloat on 20 August and two days later the vessel was in Calliope dry dock for further police inspection. The more thorough inspection revealed the extent of the damage:

> The engine room hole was eight feet long by six feet high with steel petalled three–four feet inboard. The engine room was utterly shattered, with piping and cables tangled. The stern casing had four fractures and was twisted to port. The rudder plating was fractured and collapsed inward. Hull plating at the stern was indented and split in many places. The propeller boss was fractured and distorted and its shaft bent. Combined with aged corrosion, the vessel was a fragile sieve. The Navy had every reason to be proud of its achievement in raising the wreck in one piece in the time that they did.[17]

Fittingly, the old ship ended its life as a fish reef at the suggestion of the president of the New Zealand Underwater Archaeological Association. 'A decent burial at sea is the only honourable end for this ship that has served so well on so many campaigns for so many years,' David McTaggart agreed.[18] After haggling over a site, while the government held the *Warrior* as evidence against the French while compensation was being sought, the Cavalli Islands were chosen as a burial place. The masts, the first major items removed in the salvage, were sent for display outside the Dargaville Museum.

On 12 December 1987, escorted by scores of small craft, the *Rainbow Warrior* was towed out to the Cavalli Islands for scuttling off Motutapere Island. There it remains, playing host to small fish and other marine life. In 2004 divers reported 'a smorgasbord of encrusting life that transforms this steel structure into a beautiful living reef', grazed on by red moki, snapper, leatherjacket and John dory while trevally and kingfish cruised just above.[19] Greenpeace used some of the French government's compensation money to buy and convert a bigger, better *Rainbow Warrior,* which has since visited New Zealand.

The French government provided a final, ironic note to the story. 'A year later a letter arrived at the Ministry of Transport head office from the United Nations containing a complaint from the French government,' G.C. Wright recalled. 'The French were objecting to the New Zealand government allowing the sea to be polluted by sinking the *Rainbow Warrior* as a diving attraction off Northland. The letter was filed'.[20]

10
Casualties of war

Distance, often blamed for isolating New Zealand and for increasing the cost of trade, has nevertheless shielded the country from external attack. Although we have demonstrated a curious enthusiasm for joining other people's wars, distance has prevented the countries we attacked from invading or occupying us. As a consequence, very few ships have been sunk in our waters through hostile action.

German raiders off the coast

New Zealand found itself at war in 1914 against Germany, Austria-Hungary and later the Ottoman Empire. Initially, the country saw few signs of action locally. Britain requisitioned many New Zealand deep-sea ships, several of which were sunk in northern hemisphere waters, but it was not until 1917 that an enemy ship — the German auxiliary cruiser SMS *Wolf* — made its presence felt off our coasts.

The *Wolf* was a converted merchant ship. Capable of 13 knots if pressed, it carried a mixed armament of medium-calibre guns (including two 5.9-inchers), mines, and a seaplane. Its first local victim was the Union Company freighter *Wairuna*, caught with the American schooner *Winslow* off the Kermadecs. On 2 June the *Wairuna* — bound from Auckland to San Francisco with a mixed cargo — sailed past the anchored raider but surrendered after the German ship's seaplane fired a warning shot. The freighter was sunk two weeks later after the *Wolf* had transferred stores and prisoners. The *Winslow* was sunk on the spot.

The *Wolf* left New Zealand waters, but had already sown minefields that would claim two important victims. The first was the 4700-ton Commonwealth & Dominion Line steamer *Port Kembla* which, off Cape Farewell at 1300 hours on 18 September 1917, hit a mine and started to list to starboard. The blast smashed the compass, carried away the wireless aerials and scattered deck cargo. The *Port Kembla* sank in less than half an hour, though thankfully no lives were lost, and the passing collier *Regulus* took the crew to Nelson.

The toll was heavy a year later when Huddart Parker's liner *Wimmera* hit another of the *Wolf*'s mines. On 26 June 1918 the ship, heading for

OPPOSITE *Niagara* survivors under sail after the sinking.
Ref: PAColl-8634, Alexander Turnbull Library, Wellington, NZ

The 'SS *Port Kembla*' stamp from New Zealand Post's '1917: The Darkest Hour' commemorative series. The 4700-ton *Port Kembla* was the first victim of the minefield the German auxiliary cruiser SMS *Wolf* had laid a few months earlier. Wartime censoring, however, told the New Zealand public that an engine room explosion was responsible, ensuring they remained unaware of how close the war had come to their shores.

New Zealand Post

OPPOSITE, TOP Survivors from the *Port Kembla* posed for a photograph at Nelson in 1917.
Ref: 1/2-015919-G, Alexander Turnbull Library, Wellington, NZ

OPPOSITE, BELOW The Union Company's magnificent 13,000-ton trans-Pacific liner *Niagara* putting on speed. The *Niagara* had both an illustrious and infamous career. Launched in 1912 to ply its trade at ports from Australia and New Zealand to the United States, the nickname 'Titanic of the Pacific' was dropped after the sinking of the real *Titanic* in favour of 'Queen of the Pacific'. In 1918 the *Niagara* was also rumoured to have been the source of the second wave of the 1918 influenza pandemic arriving in New Zealand, however, the official 1919 Health Department report on the pandemic concluded that the *Niagara* carried nothing more than 'ordinary influenza'.
Ref: 1/2-015286-G, Alexander Turnbull Library, Wellington, NZ

Sydney from Auckland, was passing Cape Maria Van Diemen in an area already thought to be mined, when a terrific explosion lifted its stern out of the water. The *Wimmera* sank rapidly. Two sleepy-eyed returned soldiers thought they were back in the trenches: 'it's all right old man, it's only a small attack', one groggily said to a cabin-mate.[1] A 90-year old woman initially tried to stay aboard to offer her place in the boats to a younger woman, but was put aboard and made it ashore safely. Twenty-six of the 151 (76 passengers and 75 crew) aboard were not so lucky, including Captain H.J. Kells, who decided to go down with his ship.[2]

In 1934 German naval strategists, struggling to rebuild the navy after its defeat and scuttling at the end of the war, started planning a new generation of commerce raiders to supplement their small surface fleet. By 1939 Germany had a pool of fast freighters and banana boats suitable for conversion into fast raiders.[3]

The first of these unwelcome visitors was the *Orion*. Carrying six 5.9-inch guns, several lighter pieces, torpedoes and mines as well as a spotter aircraft, the *Orion* entered the Hauraki Gulf on the night of 13 June 1940 to lay a minefield across the approaches to New Zealand's busiest seaport. For ten nerve-wracking hours, always in sight of the mainland, and with the light cruiser HMS *Achilles* and the armed merchant cruiser HMS *Hector* within striking range, Commander Kurt Weyher laid 228 mines in zigzagging barrages, one across the eastern approach to the passage between Great Mercury Island and Cuvier Island and the other, more extensive one across the northern approaches to the Hauraki Gulf, running from the end of Moko Hinau Island, outside Maro Tin Island in the Hen and Chicken Islands and finishing up well within sight of the Northland coast. All were moored contact mines. Then Weyher sped away.

The mines took just five days to draw blood. Early on the morning of 19 June 1940, the realities of war hit New Zealand for the first time, when an explosion blasted the Union Company Vancouver

Shipwrecked crew of the Port Kembla at Nelson, N.Z. 18.9.17. F N Jones

A mine swept up from the Hauraki Gulf, Auckland, in the wake of the *Niagara*'s sinking. Of the 228 mines the *Orion* had originally laid, 106 would ultimately be accounted for. The others were assumed to have sunk or broken free. Within three months of the *Niagara*'s sinking the area was declared mine-free.
Ref: PAColl-5547-084, Alexander Turnbull Library, Wellington, NZ

mail liner *Niagara* apart.[4] The *Niagara* had been a household name in New Zealand since 1913. When built, the 13,415-ton liner was described as 'the biggest and most luxurious liner that has ever come south of the line'. Its oil-burning machinery was innovative, and in 1940 the *Niagara* was still highly regarded.

The *Niagara,* too, had been spotted by the raider *Wolf*'s plane in World War I, but had been too fast to be caught. This time there was no reprieve. Voyage 163 was to have taken the ship from Auckland to Vancouver via Fiji. Aboard were 136 passengers, 203 crew and a cargo that included half the New Zealand army's small arms ammunition, urgently needed for Britain's defence. Also tucked away were 590 bars of Bank of England gold. Valued at £2,500,000 (worth nearly NZ$120 million in present-day terms), the gold was to pay for arms bought from the United States. Without knowing it, Weyher was right on the money.

At 0340 hours in the fairway between Bream Head and Mako Hinau Island, a terrific explosion near the forward hold threw passengers from their bunks and brought the big liner to a complete halt. AB F.D. Harris, walking towards the No. 2 hatch, heard 'merely a muffled blast', but was bowled over by the blast, which temporarily blinded him.[5] High up in the crowsnest, AB Reginald Kerr heard the explosion and looked down 'and could see all the hatches and beams blown off the No. 2 hold and grey smoke was coming out of the hatch with a strong smell of cordite'.[6] On the bridge, the quartermaster, William Hart was thrown backwards from the wheel, hitting his head by the door of the D.F. cabinet. 'All the wheel-house windows were blown in, the helm seemed to be rendered useless.'[7]

Captain William Martin was woken by the explosion 'almost under my quarters ... My room was more or less wrecked.' After slipping on his trousers and overcoat, Martin rushed to the bridge where he discovered that his ship was settling by the head. 'Looking down Number 2 hatch the lower hold appeared to be full of water with debris floating about and beams and portions of hatches scattered around.'[8] He ordered action stations, which covered the closing of all watertight doors, and swinging out the boats. But he knew his ship was doomed. While his radio operator broadcast distress messages and the crew started firing rockets, he gave the order to abandon ship.

Fortunately the night was calm and clear, the ship's internal lights stayed on, and the passengers boarded the 18 boats without too much difficulty. No one had sustained any serious injury, the only fatality being Aussie, the ship's cat, who, not finding the lifeboat to his liking, scrambled back aboard the liner to go down with it. At 0532 hours the passengers and crew aboard the boats were able to watch the *Niagara,* which had been sinking by the bow, slip beneath the waves.

Help was at hand within about five hours, led by the RNZAF's high-speed rescue launch. Fear of mines kept the bigger ships such as the

Wanganella and the *Achilles* at a respectful distance, but the coaster *Kapiti* and boats from the *Achilles* quickly picked up the passengers who were none the worse for wear despite their involuntary messing about in small boats. By evening they were back in Auckland and minesweepers were tackling the first mines. The real victims were the Gulf's penguins, which suffered many fatalities from oil slicks.[9]

Salvaging the gold was a story in itself. The Bank of England hired United Salvage Proprietary Ltd of Melbourne which acquired the old coaster *Claymore* and set up operations from Whangarei in December 1940. By 2 February 1941 divers had located the wreck at a depth of 134 metres. This was deeper than anything previously worked on (the record was 120 metres, held by Italian salvors on a wreck off Ushant), but the huge amount at stake spurred on United's divers. Lying on its side at an angle of 70 degrees, the *Niagara* would have to be blasted apart to provide access to the strongroom.

It was not easy. The *Claymore* had to lie buoyed in the middle of a still largely unswept minefield (two mines bobbed up during salvage operations) and was frequently buffeted by heavy seas. Things were little calmer for the men working in the diving bell. It bounced up and down, flung about whenever the sea hit the *Claymore* until United's men solved the problem by tensioning the winch to maintain position. It was a tough, unpleasant job, conducted under conditions of wartime secrecy. The divers broke into the strong room on 13 October and brought up the first two bars of gold. By 7 December when salvage operations ceased, almost 95 percent of the gold had been recovered.

The remaining 35 bars still enticed others. Using new technology, the *Foremost 17* recovered 30 in 1953 over a three-month period. The others will probably remain with the wreck.

The *Orion* had not, however, finished with New Zealand. Its second mine victim and, for New Zealand, the last loss in home waters through hostile action, was the minesweeper HMS

NIAGARA

More than 60 years after its sinking by a German mine, the *Niagara* continues to make the headlines. In 1999 concerns were expressed about small quantities of oil leaking from the wreck's corroding bunkers. That year, too, the ship featured in an episode of the television series 'Shipwreck' and there have been two books on the ship and the salvage of the gold. At more than 120 metres, the wreck is too deep for most recreational divers, but in recent years several expeditions have been made to film and explore the wreck, some using ROVs. In 1999 Tim Cashman reported that the wreck was still impressive. The decks above the first level of superstructure had collapsed, but the fore mast was still upright, 'now pointing horizontally east towards the Mokohinau Islands ... black coral trees grew off it at various points and the crows nest still afforded a panoramic view over the forward deck'. Further aft, the team saw 'the strongroom crater', the section of the ship blasted apart by the salvors all those decades ago. 'The carnage was incredible ... twisted girders, sheets of steel plating, rust scale, pipes and other debris littered the area ... It would take forever to sift through the debris to find any gold bars amongst that lot. No wonder the salvors classed further work as unviable!'[10] But, such is the romance of the gold story, it is likely that further interest will be shown in the ship and its elusive bullion.

Puriri. Pride of the Anchor Shipping & Foundry Company, the *Puriri* had been a naval vessel for less than a month. On 14 May 1941 the officers and crew were still getting used to their newly converted ship while working up with the other ships of the 25th Minesweeping Flotilla, when they were sent to sweep the minefield in which the *Claymore* was still working.

The launch *Rawea* had buoyed a mine the previous day, but the *Puriri*'s watch-keepers did not see it in time. A huge explosion rocked the vessel, which sank within minutes. With it went commanding officer Lieutenant D.W. Blacklaws, petty officer B.A. Mattson, stewards G.E.R. Hobley and J. Richardson and able seaman L. Purkin.

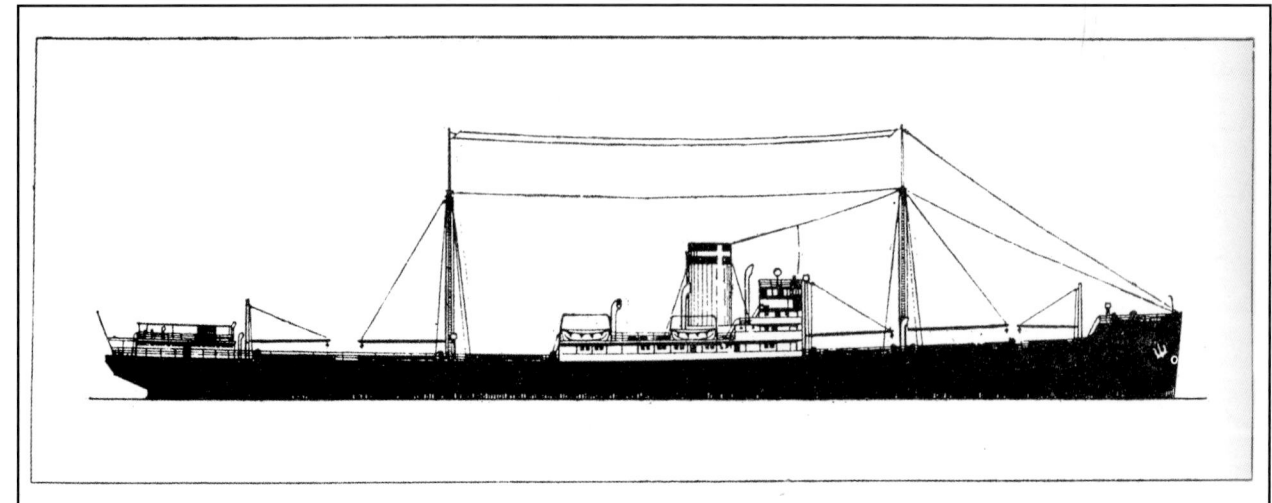

THE RAIDER ORION (SHIP NO. 36)

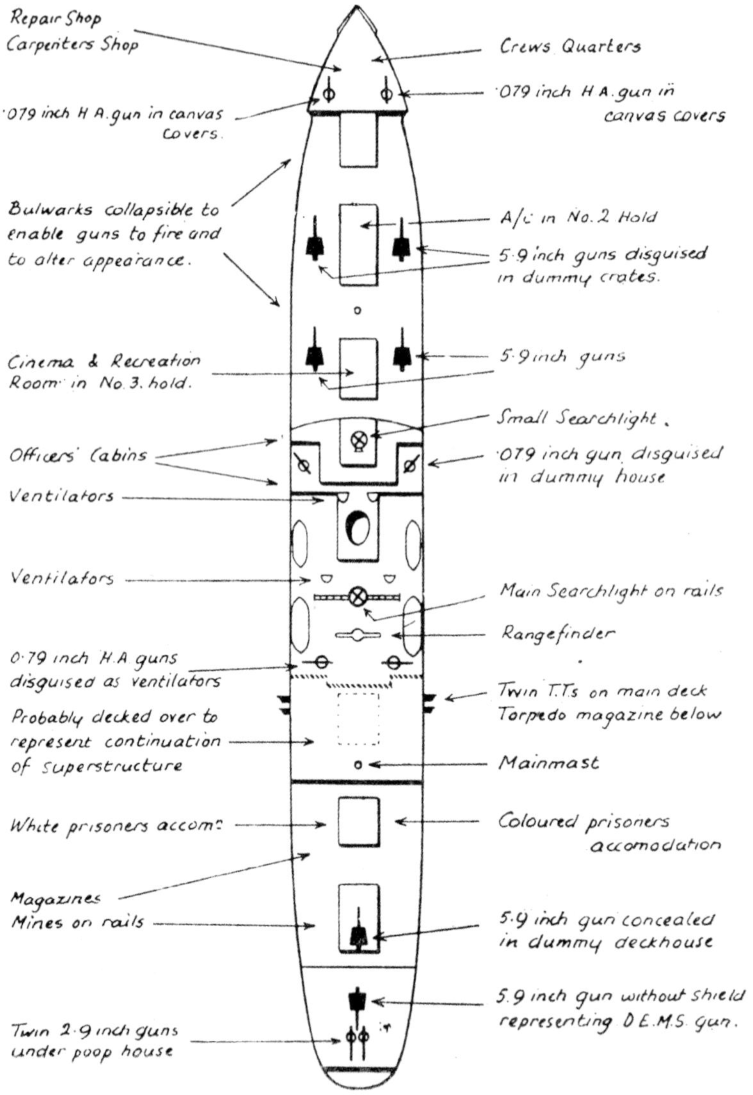

DECK PLAN OF THE RAIDER ORION, as in May 1941

Labels (clockwise/around plan):
- Repair Shop, Carpenters Shop
- Crews Quarters
- .079 inch H.A. gun in canvas covers
- .079 inch H.A. gun in canvas covers
- Bulwarks collapsible to enable guns to fire and to alter appearance.
- A/c in No. 2 Hold
- 5.9 inch guns disguised in dummy crates.
- Cinema & Recreation Room in No. 3 hold.
- 5.9 inch guns
- Small Searchlight
- Officers' Cabins
- .079 inch gun disguised in dummy house
- Ventilators
- Ventilators
- Main Searchlight on rails
- Rangefinder
- 0.79 inch H.A. guns disguised as ventilators
- Twin T.Ts on main deck Torpedo magazine below
- Probably decked over to represent continuation of superstructure
- Mainmast
- White prisoners accom:
- Coloured prisoners accomodation
- Magazines, Mines on rails
- 5.9 inch gun concealed in dummy deckhouse
- 5.9 inch gun without shield representing D.E.M.S. gun.
- Twin 2.9 inch guns under poop house

HMS *Gale* picked up the 26 survivors, five of whom were wounded.

The rest of the 25th continued their dangerous and unglamorous task and by the end of September had accounted for 106 of the 228 mines originally laid. Since it was assumed that most of the others had broken loose or sunk, the area could now be declared mine-free.

After heading up into the Pacific for a while, the *Orion* returned to New Zealand waters a few months later. On the evening of 20 August 1940, the raider was cutting through the Tasman between Cape Egmont and Wellington when it sighted the next victim rolling in the heavy seas, the New Zealand Shipping Company's freighter *Turakina*. The ship was heading from Sydney to Wellington to top up with frozen meat.

One of a series of similar, solid-looking, single-funnel coal-burners turned out for the NZSCo and Federal in the early 1920s, the *Turakina* was no flyer. It was really just a huge floating freezer unit — and a poorly armed one at that — sporting nothing but one vintage 120-mm gun bolted to the poop manned by a small contingent of naval gunners. But Captain J.B. 'Jock' Laird was no quitter and is reported to have vowed before leaving Sydney 'that if attacked his ship would fight to the end'.[11] He may also have been inspired by the memory of another NZSCo vessel, the *Otaki* against the *Moewe* in World War I. In any case, Laird decided to fight it out, hoping to delay the raider long enough for warships to catch it. Ignoring the raider's requests to stop and maintain radio silence, the *Turakina* put on speed, turned away from the *Orion* and radioed that it was being shelled by an enemy raider.

The *Turakina* never stood a chance. The heavily armed *Orion* scored with the first salvos and quickly reduced the *Turakina* to a blazing wreck. Down went the foremast, together with the lookout, and then the engines failed. Within 20 minutes the freighter was sinking by the stern and half the 56 crew lay dead or wounded. Shooting blind after the early loss of their

OPPOSITE A deck plan of the *Orion*, showing the placement of the concealed weapons.
New Zealand in the Second World War, Episodes & Studies 'German Raiders in the Pacific'

BELOW In August 1940 the New Zealand Shipping Company's *Turakina* put up a gallant fight in the Tasman against the German raider KMS *Orion*. After the war the Oamaru friends of Captain J. B. Laird erected a memorial to this courageous mariner on their waterfront.
Gift of Wellington Museums Trust, New Zealand Maritime Museum (2012.0.4850)

rangefinder, the *Turakina*'s gunners had the satisfaction of straddling the German ship on a number of occasions, bending some hull plates and sending splinters whizzing along the vessel's decks. Even then Laird refused to give up, racing astern through the flames to urge his surviving gunners to 'have another shot at the —!'

These shots, part of the only gun duel ever fought in the Tasman, kept Weyher at a distance and encouraged him to expend two precious torpedoes on the troublemaker. Both hit, one astern with little effect, the second amidships, where it completed the work of destruction begun by the gunners. At 1822 hours the *Turakina*'s survivors abandoned ship, leaving 35 dead on its blazing decks. Despite the risk of detection, Weyher picked up his gallant adversaries. The *Orion* had travelled a long way to reach these happy hunting grounds, so it was perhaps unrealistic to expect that Wehyer would abandon them. Avoiding Allied ships and aircraft, he crossed the Tasman before rendezvousing with the raider *Komet* and the supply ship *Kulmerland*. By the morning of 25 November, they were back off the New Zealand coastline, searching for new victims, when smoke was seen on the horizon.

This was an unlucky little ship, the *Holmwood*. Built at Goole in 1911 and better known to New Zealanders as the *Tees*, the 546-ton coal-burner had been acquired a few months earlier by Sydney Holm, renamed *Holmwood* and fitted out to supply the Chatham Islands. Once a month it plodded slowly from Lyttelton with a few passengers and a mixed cargo.

On the morning of 25 November, the *Holmwood* was on only its second trip back for Holm, carrying a dozen passengers and a cargo that included a horse and 1375 live sheep, when chief engineer Fred Abernethy sighted a ship bearing Japanese markings. Japan was then still neutral. Captain James Miller altered course to put the mystery ship astern, as advised by navy procedures, and ordered the passengers to dress and the boats swung out.

They were still studying the ship which was now following them when they were surprised by the appearance of two more vessels. At that point the *Orion* hoisted the Kriegsmarine flag and instructed the *Holmwood* 'for the benefit of your company you should stop immediately'. As the raider was end-on, the *Holmwood* bridge crew failed to read the signal flags and a warning shot had to be fired before its crew understood the morse signal 'Stop immediately'.[12]

Captain Miller felt he had no choice. Unarmed and even slower than the *Turakina,* the *Holmwood* was not designed for heroics. Since atmospheric conditions made radio reception and transmission poor, he stopped engines at 0800 hours and waited for a German boat to come alongside. There the Germans, who had planned to transfer the crew to the boats, discovered that there were 12 passengers in addition to 17 crew. They decided to transfer them to their ship. A prize crew replaced the red ensign with the swastika, restarted the *Holmwood*'s engines and steamed it a safe distance from the Chathams. At about 1300 hours 14 shots finished off the little ship.

Long before then, the authorities knew that something was amiss. On 2 December the *Dominion* reported that the *Holmwood* was overdue. Twelve days later the Chathams' policeman, Constable C.L. Spencer, reported finding a hatch beam near Kaingaroa 'pitted with shrapnel'. Two days later the tide brought in pieces of timber similar to the ship's bridge.[13]

The navy, after debriefing the crew in January, was unimpressed with Miller's behaviour. The chief of naval staff criticised Holm Shipping for sending the ship to sea with just one radio operator, Miller. He also dismissed Miller's reasons for not broadcasting as 'a lame one ... There was every chance that his transmission would be picked up'. The transmission would have saved the next victim, the liner *Rangitane*.[14]

The New Zealand Shipping Company's *Rangitane* had left Auckland the day the Germans intercepted the *Holmwood*. One of three

The raider victim *Holmwood*, seen here under its old name, *Tees*.
Gift of Wellington Museums Trust, New Zealand Maritime Museum (2012.0.5030)

pioneering motor liners on the New Zealand-Britain run (the others were the *Rangitiki* and the *Rangitata* the 16,737-ton, 17-knot *Rangitane* was one of the most prestigious ships serving New Zealand. Aboard were 11 passengers, 200 crew and a full cargo of valuable primary produce.[15] The ship's high speed enabled it to sail unescorted. At 0252 hours on the 27th, while the *Rangitane* was approximately 300 miles off East Cape, the first officer sighted suspicious vessels. Called to the bridge, Captain H.L. Upton ordered the radio operator to broadcast a warning, drawing down a nasty cross-fire from the *Orion* and *Komet*. Upton's ship could have outrun either ship but, worried about his passengers and the accuracy of the German gunnery, he stopped once he was satisfied that the radio message had gone through.

The *Rangitane* had taken quite a beating. F.F. Howells, its chief butcher, recalled waking to the sound of alarm bells. As he made his way to the boat deck, F and E decks seemed quite normal, but D deck was full of smoke. He and three shipmates helped four people they found lying on the deck, including a stewardess.[16] Another crewman, L.N. Sowerby, looked out before going to his station and was 'dazzled by a powerful searchlight'.[17]

The *Orion* and the *Komet* continued shooting for some minutes, eventually killing five passengers and five crew and wounding several others, one of whom later died. When German boarding parties clambered on to the biggest ship ever sunk by an armed raider, they found it ablaze on all upper decks. They opened the *Rangitane*'s seacocks and fired several more salvos before heading away at speed for the Pacific where, at Nauru, they accounted for several more Allied ships, including the Union Company's new freighter *Komata*. The captured passengers and crew were left on Emirau Island later that year.

PART FOUR
The Modern Era

ABOVE The *Star of Canada* was only 30 months old and was one of the latest refrigerated ships when she came ashore on the Kaiti rocks, Gisborne, on the stormy night of 23 June 1912. While lying at anchor she was struck by a squall, which broke her bow anchor cable. Within minutes the *Star* was on the rocks. Salvors worked furiously for 10 days to save the vessel, pumping out the forehold and removing cargo to lighten the ship, but to no avail. Stripped and abandoned, the ship settled lower and lower in the water, eventually disappearing completely.
Tairawhiti Museum

PAGES 146–47 The MV *Rena* suffered grounding on Astrolabe Reef in October 2011. On 8 January 2012 her stern began sliding off the reef. The vessel's breaking up had for some time been seen as inevitable, though the timeline was anybody's guess. It happened on the night of 7–8 January, when a severe storm battered the fragile wreck with six-metre waves. By daybreak the two halves were 20–30 metres apart and the stern was slowly beginning to slide off the reef. Within two days three-quarters of the *Rena*'s stern section was underwater, the remainder not disappearing until 4 April.
New Zealand Defence Force

11

Conference Lines casualties

From the 1890s the number of Conference Lines ships (a reference to the British monopoly that dominated New Zealand's export trade through its price-setting 'conference') lost here declined rapidly. The reasons were many, but the principal ones were the use of larger, more powerful ships, safer, better ports, better-lit and charted coasts, and better masters.

Even so, human error or bad weather still claimed scalps. Between 1912 and 1939 several big, well-found ships — the *Star of Canada, Devon, Tyrone, Tongariro, Port Elliott, Wiltshire* and *Port Bowen* — were wrecked in New Zealand waters, usually at night and in bad weather. This chapter will look at some of these ships.

As any seafarer knows, the hours either side of midnight are the most dangerous when the human body is at its least alert and visibility is poor. Late on the night of 23 June 1912 the 4623-ton *Star of Canada* was anchored in the Gisborne roadstead when the weather, already bad, turned into one of the wildest southerly squalls ever experienced locally. At 2200 hours Captain J.M. Hart, knowing the dangers of the anchorage, decided to head for sea. Unfortunately, the boilers had been shut down for cleaning, so it would take time to get moving. Too long, as it turned out. At 2315, while the machinery was being readied, the ship started to drag its anchor and drift towards the shore. Before a second anchor could be dropped the *Star of Canada* struck a rock and started flooding. At 2330 Hart fired distress rockets but with the port's small tug *Hipi* too puny to help, he dropped his other anchor and tried to keep the ship bow-on to the waves.

By the next morning the *Star of Canada* lay on the seabed with its three forward holds flooded. During the night Hart had lost the struggle to keep the ship's bows pointing to the seas, so now it lay almost broadside to the waves, and was taking a terrible pounding. Broken water fore and aft indicated that the ship was trapped in a circle of jagged rocks. The *Star of Canada* was going nowhere and broke its back a fortnight later. The deckhouse was lopped off and deposited on a section in Gisborne (and transferred to the Tairawhiti Museum in the 1980s) and some cargo was recovered, but the rest of the wreck sank, although parts of its hull still provide interest to divers. The inquiry exonerated Hart, blaming the accident on the unpredictably sudden and furious nature of the gale.

> **Close to the entrance, the sea can race through the narrows very fast in bad weather. And that night's weather was atrocious.**

'An awful heave and crash' — the *Devon*

'Rocks on the starboard bow, sir! Rocks on the starboard bow!' It was the last thing Captain Arthur Caunce wanted to hear, but by the time he had ordered 'starboard your helm!' it was too late for the *Devon*. A howling southerly was lashing Wellington that August 1913 night, driving thick rain before it as the 5489-ton Federal-Houlder-Shire liner started its run into port. Heavy seas had been battering the ship for days and Captain Caunce was looking forward to a snug berth at the capital's wharves.

It was not to be. At 2015 on 25 August, the *Devon* went ashore near the lower Pencarrow Head light and became a total wreck. Here, close to the entrance, the sea can race through the narrows very fast in bad weather. And that night's weather was atrocious. Captain A.M. Edwin, commanding the inter-island ferry *Wahine* whose bows were dipping well into the water, reported poor visibility, 'most violent bursts of wind, short, sharp and of great power, straight from the Antarctic, and bitterly cold'.[1] Although not the worst southerly he had encountered, it was bad.

The *Devon* hit with a sharp, rending sound that caused Caunce to stop the engines and blow the boilers. Water rushed into the stokehold, which, along with the engine room, flooded in two minutes. Engineers and firemen scrambled for their lives but must have been dismayed by what they saw above decks: waves were breaking over the stern, jolting the ship badly. Caunce mustered the crew forward and fired rockets.

The lighthouse keeper had already alerted the authorities. Unfortunately, the storm prevented Caunce from launching lifeboats or rescue vessels from getting close. The ferry *Duchess* tried but failed. The tug *Karaka* sailed to help but was forced back. While the *Devon*'s seamen huddled on the crowded forecastle, and the *Wahine* stood by impotently, Fort Dorset's searchlights played across the harbour, providing a psychological boost if nothing else.

Although the seamen could hear the shouts of people on the beach, the wind thwarted their efforts to get a line ashore. After many exhausting failures, they decided to wait for dawn. Daybreak brought some encouraging signs. The wind had weakened slightly, and a small party of about 20 people, led by harbourmaster Captain J. Johnson, had assembled on the beach. A reporter thought that the *Devon*'s crew, shivering miserably on the forecastle, had been very lucky. 'It was a piece of good fortune that the body of the vessel was gripped strongly by the rocks,' he wrote with an optimism the sailors may not have shared. 'If the steamer had swung around broadside after striking all lives might have been lost last night. Just about the bow rocks jut up like giant teeth, whetted for destruction, and flecked with the foam of ravening appetite. They would have pierced, cut, and crunched the labouring ship and how could men have lived in those wild waters in darkness?'[2]

The crew resumed its efforts to get a line ashore. Twice life-buoys attached to ropes either slipped off or got hopelessly entangled in rocks far beyond the reach of the shore party. A third time, however, a line attached to a lifebelt was plucked out of the heaving seas and brought ashore by rescuers, some of whom were almost dragged under by the treacherous backwash. 'For some hours they proceeded, but the work was extremely difficult', the *Evening Post* reported. 'Then one of the bulkheads was pressed into service and, making their way from rock to rock, some of the party at last got close to where the wires were entangled amongst the pinnacles. To get within reach, however, it was necessary to place a plank between two jagged outstanding rocks over which the sea was breaking perilously. This dangerous task was safely accomplished; however, the feat being breathlessly watched by those on ship and shore alike.'[3]

Even the press helped with the long, laborious job of hauling the crew ashore one by one. 'And with a "Haul away, boys!"', the swaying basket made its jerky, bouncy way along the hawser. First to reach dry land was the quartermaster, who made heavy going of the three-minute journey, being severely buffeted by the waves: 'as he neared the "sag" of the hawser a big comber came curling towards him, swirled round him and drenched him to the waist. Another haul on the shore, and the dripping mariner was extricated from his basket and helped ashore. He was not alone,' the *Dominion* reported. 'Shivering in his arms was a small kitten, which mewed plaintively when put down beside the fire.'[4] Last ashore were the chief engineer and Captain Caunce, each weighing over 96 kg. They landed to wild applause. An impromptu celebratory meal brought out by

The *Devon* was one of many ships lost near Pencarrow Heads at the entrance to Wellington Harbour. Note the lines running between ship and shore. Salvage assessments were undertaken, but alas, there was nothing that could be done for the *Devon*.
Ref: APG-1361-1/2-G, Alexander Turnbull Library, Wellington, NZ

packhorse followed. It included 'an ample supply of stimulants to warm the 14 chilled and wet mariners,' since the shipping company had sent three cases of whisky to the wreck site.[5]

The aftermath of the *Devon* accident was messier than normal. The court of inquiry, in a majority decision, suspended Caunce's certificate for three months, blaming him for taking the *Devon* too close to Pencarrow Head and too fast, and for mistaking the lights.[6] But Caunce appealed, claiming a breach of justice and of ethics through making Captain Felix Black an assessor on the court. Caunce had earlier declined Black's offer to pilot his ship from Auckland to Wellington and believed that that may have biased Black against him; in any case, Caunce felt that coastal pilots should not serve on inquiries because they had a pecuniary interest in increasing the use of coastal pilotage. The minister of marine rejected this argument, but changed his mind when the solicitor-general advised that Caunce's objections were 'well founded'.[7]

'The Second Devon Finding — Complete Reversal Captain Blameless — Certificate Returned — Port Lights Impeached' — the headlines said it all.[8] The second inquiry completely overruled the magisterial inquiry, finding that Caunce had mistaken the Falcon Shoal buoy for the red sector of Somes Island light, an error — but not one that made him responsible for the accident, which it blamed on the weather. For the *Devon* there was no second chance. It quickly disintegrated, scattering cargo and fittings along the shore.

What a difference a fine day makes! This photograph, with the wrecked ship so close to the Pencarrow lighthouse, shows how bad visibility must have been when the *Devon* went ashore.
Ref: PAColl-9004, Alexander Turnbull Library, Wellington, NZ

> Salvaging the cargo kept Richardson & Co. busy for weeks. Manager Kenneth McLeay took an oil winch out to the wreck to load his small craft. One of the ship's holds was filled with whisky and brandy and this, naturally, received priority.

'TAKE THE SEATOUN CAR'

Wrecks sometimes provided windfalls for the enterprising. In the case of the *Devon*, which was large and highly visible, the first to cash in were the harbour ferry operators and charter boat owners, who took boatloads of sightseers in the days following the wreck. So concerned about looting were the authorities, that they asked the lighthouse keepers to keep an eye on the registration numbers of small fishing boats approaching the scene. But it was not even necessary to take to the water to get an eyeful. News spread that good views of the *Devon* could be obtained from Breaker Bay, and posters quickly appeared on city streets advising the public to 'Take the Seatoun Car' — the tram.[9] In August the *Dominion* said: 'A report was received in town yesterday that a certain kinematographic picture [movie] firm was busy securing moving pictures on the beach yesterday afternoon. On inquiries being made it was found that this firm's operators were very energetic in getting a representation of the real thing to put before the public,' the paper noted. 'They had carefully undone the lashings which had secured the life basket and line to the wreck and when the chief officer arrived on the scene members of the party were industriously pulling the life basket containing a man up and down in front of the camera. Naturally the officer stopped this, but a moving picture had been taken all the same.'[10]

The *Tongariro*'s troubles

On the evening of 30 August 1916, the New Zealand Shipping Company's steamer *Tongariro* was about halfway through its voyage from Auckland to Wellington when Captain Harry Makepiece realised — far too late — that something was wrong. With a resounding crash, the big ship struck Bull Rock off Portland Island (between Gisborne and Napier) with its stern, swung around and came to a halt, listing badly to starboard and with its bow on the rock. It was immediately obvious the ship was doomed. It had been steaming fast and broke in two on impact, shattering right across the No. 3 hold.

Fortunately, none of the 96 officers and men was seriously injured. They made for the boats, many in just their night attire. The ship's list made it hard to launch the boats, and five men were slightly injured when a port-side boat (the starboard-side ones could not be launched because of the list) was smashed, throwing its occupants into the water. There was no panic and most of the crew was picked up three hours later by the NZSCo's lighter *Koutunui*, and transferred to the liner *Westralia*.

Meanwhile, settlers on the Mahia Peninsula spread news of the disaster. By the time the company's local manager reached the site, the *Tongariro* was still afloat, its bow supported by Bull Rock, but all the holds and the engine room were waterlogged. The decks were above water but seas were breaking over the bow, and it could not be long before the *Tongariro* slipped off Bull Rock and sank in the deep water under its stern.

Salvaging the cargo kept Richardson & Co. busy for weeks. Manager Kenneth McLeay took an oil winch out to the wreck to load his small craft. One of the ship's holds was filled with whisky and brandy and this, naturally, received priority. Richardsons' men recovered much of it by smashing one case in each tier to free it, but when the *Tongariro* eventually broke up it was 'whisky galore' as locals engaged in some very earnest beachcombing. Some outwitted Customs patrols by burying their windfalls for recovery once the authorities had tired of the game.

The court of inquiry censured the master and the chief officer, finding that the mishap was primarily caused by the latter failing to take ordinary precautions to verify his position when off Table Cape. Both had their certificates suspended. The *Tongariro*, a steel, twin-screw steamer of 8073 tons had been built in 1901.

The *Wiltshire* hits an island

One of the more dramatic interwar era wrecks occurred on remote Great Barrier Island on 31 May 1922, when the stately five-master Federal liner *Wiltshire* piled up on the rocks at remote Rosalie Bay, near the southern tip of the island. Just 10 years old, the *Wiltshire* was one of the biggest ships trading to New Zealand. The ship had been punching through filthy weather. Heavy wind and torrential rain had lashed it for half a day and visibility had fallen to less than a ship's length. Suddenly, there was a terrific crash, then four heavy bumps as the steel monster forced itself up on to the rocks. Seconds before, hearing a sailor cry out that the water was shallowing, Captain Bertram Hayward put the helm aport to turn around but 'she only turned a little bit before she struck the shore'.[11]

Captain Hayward summoned the 102 officers and men topside but decided not to launch the boats. With heavy seas breaking over the *Wiltshire*

RIGHT, TOP The *Wiltshire* seen through trees at Rosalie Bay; the ship's back is obviously broken and its stern has already slipped beneath the waves.
Auckland Libraries Heritage Collections 7-A9909

RIGHT, BELOW The *Wiltshire* provided dramatic challengers to rescuers. Here two junior radio operators can be seen being brought ashore on a 'bosun's chair' rigged up by Auckland naval personnel.
New Zealand Herald

it would have been sheer folly — they would either be dashed against the ship or smashed apart on the rocks off the cliff face they could see ahead of the bow. So the men settled down to a nervous night in the saloon and prayed for better weather at day light.

But things only got worse. At 1130 on the 1st, the *Wiltshire* snapped in two just abaft the No. 4 hatch with a deafening bang. Hayward leaped from the bridge down on to the deck, followed by his officers, who used a rope to join the crew on the forecastle. The *Wiltshire* was a complete wreck. The ship's bow section was firmly impaled, but its stern jutted out over deeper water and was beginning to settle. Would the ship snap in two before the crew could escape?

By now the coasters *Katoa* and *Arahura* had arrived. While the *Arahura* stood by and maintained radio contact with Auckland, the *Katoa* landed rescue parties, storm gear and blankets. They teamed up with islanders who had kept watch all night. 'Heartrending was the experience of watchers on the cliffs above the wreck prior to the line coming ashore,' a reporter recalled. 'The watchers could hear cries from the ship but could do nothing. Hope was raised when one of the crew attempted to swim ashore with a line but this died down when the enormous seas running caused the effort to be abandoned, and the exhausted swimmer had to be hauled on board again.'[12]

For Aucklanders following the breaking story, it was a tense time as the news varied between optimism and deepest pessimism. At 1345 the *Katoa* radioed Auckland that the *Wiltshire* had broken in half, the stern had disappeared and that the bow was crowded with people. That set Auckland abuzz. News had first reached the city by radio late the previous night and had been all bad; the picture of trapped crew huddling on the bow daunted everyone. Next day someone suggested using a seaplane to get a rope aboard — an early instance of air-sea rescue techniques — but the attempt had to be abandoned.[13] The gloom was quickly compounded by news from the Coromandel that the crew was considered beyond help. The seesawing spirits got another boost less than an hour later, though, when it was reported from Tryphena that a line had been got ashore and that four men had already landed. 'Prospects of rescue now good.'[14] Two hours later the *Katoa* reported that the line still held and that the 99 men on the wreck were well sheltered, well fed and waiting for daylight when a party from HMS *Philomel* planned to set up a breeches buoy. With the seas moderating slightly and the forward part of the ship showing no sign of shifting, Hayward had decided to wait for a safer means of getting ashore.

The hero of the hour was the 24-year old seaman from the *Katoa,* Wilfred Kehoe. He scrambled down the dangerous cliff face and grabbed a line attached to a hatch cover that had been floated ashore. He then made the slow and dangerous ascent. After what must have been an agonising delay, a line was fastened to a tree on the cliff and the first man was hauled ashore metre-by-metre. Three others had followed before operations halted for the night.

The next day's rain did not dampen the spirits of the *Philomel* party. With 10 men at a time hauling on the rope, they brought two sailors ashore every six minutes. Captain Hayward had wanted to be the last to leave, but because of his age and infirmity, he was persuaded to go earlier 'with smiling lips disguising a breaking heart'.[15] By 2200 on the 2nd, two days after the *Wiltshire* went ashore, everyone was safe. The crew went to Auckland where a large crowd was waiting. Governor-General Lord Jellicoe offered them fresh clothing.

A few mail bags were recovered, but most of the 10,000 tonnes of general cargo was lost, along with the ship. The *Wiltshire*'s loss was serious because it was one of a few ships recently chartered to open up a direct trade with Manchester. Telescoping masts and funnel enabled the ship to navigate the Manchester Canal.

The court of inquiry found Hayward guilty of two errors of navigation: in not exercising caution when unable to pick up Cuvier light when expected, and in not acting immediately on the danger signals given by the sounding taken about 20 minutes before the ship struck. He had to pay the costs of the inquiry.

'This wreck no blurry good' — the *Port Elliott*

The hours before midnight claimed another victim just a few years later. The ship was the Commonwealth & Dominion Line's 7395-ton steamer *Port Elliott* — the place was Horoera Point, between Te Araroa and East Cape, and the date was 12 January 1924.

Better known to New Zealanders as the *Indrabarah,* the *Port Elliott* had already had one close escape, running aground on Rangitikei Beach in the Taranaki Bight in May 1913. Early newspaper reports gave the ship up for lost but the fact that it was embedded in sand, not rocks, kept its salvage chances alive. But as one salvage attempt after another failed, it gradually worked its way up the beach, closer to the waterline. Then, early in July, the combination of a very high tide, a wind from the right direction and some strong pulling from the Union Company tug *Terawhiti* freed it.

This time, however, the ship's luck ran out. The night was pitch dark and the fog was thick. The crew had no inkling of trouble until the big ship — steaming at near maximum speed (13 knots) —

The Union Company's recently acquired 'Irish County' steamer *Tyrone* made landfall the hard way at 0400 on 27 September 1913, when she ran ashore off Wahine Point at Otago Heads. It was so well insured that the company made a tidy gain on the payout.
Gavin McLean Collection

crashed into the reef, sliding over it and coming to a complete halt. Captain A.J. Fishwick ordered full speed astern, but to no avail. The ship — pinned to the reef beneath its engine room — was stuck. As water rushed into the forehold and the engine room, by the early morning the *Port Elliott* was down by the bows.

An Auckland radio station picked up the SOS message at 2352. Minutes later a Wellington station received one from the *Port Elliott* saying that it was ashore near East Cape. The Wellington station notified the sloops HMS *Laburnum* and *Veronica*, the government steamer *Tutanekai*, and the Union Company coaster *Arahura*, as well as the C&D Line's *Port Victor*. All diverted to the accident scene, followed by smaller steamers from Napier and Tolaga Bay.

Horoera pa residents had been roused from their slumbers at about 0330 hours by the shrill cry of a ship's whistle. One local boy, Norman Bryant, set out on horse for Gisborne, unaware that the ship's radio was functioning. Others gathered around, convinced that the whistle was coming from one of Richardson & Co.'s small wool boats. By daybreak the mist had lifted sufficiently for the crowd to see the *Port Elliott*, apparently little damaged, with its propellers half out of the water at half-tide. Soon everyone was in on the act. One man rushed into the East Cape lighthouse keeper's house shouting, ' A boat is sinking off the East Island; only her masts are left.'

The *Tutanekai* reached the scene at 0400 and stood by about a kilometre from the wreck. The *Port Elliott*'s crew, who had taken to the boats several hours before, sent some boats to the *Tutanekai* and some back to the ship, which was obviously in no danger of breaking up. A few hours later, however, a moderate sea started bending some hull plates.

With the arrival of the East Cape lighthouse keeper, shore-to-ship communication was at last possible; Smith, the keeper started signalling by means of handkerchiefs tied to bits of sticks! In this way arrangements were made for the crew to land on the shore in the evening and go to the nearby meeting house, where preparations had been made for them. By now, Richardson & Co. was deploying its steam lighters to the wreck site to remove the cargo, to the disgust of one Horoera Maori, who was reported to have muttered: 'This wreck no blurry good. Nothing washed ashore like the *Tongariro*. *Tongariro* wreck kapai.'[16]

The salvors did well out of the *Port Elliott*. Using the small steamers *Fanny* and *Ruru*, Richardsons offloaded a considerable amount of general cargo, cased benzine and 107 cased motor vehicles. Dried out and spruced up, they fetched reasonable prices at a Wellington auction. By 15 January the ship was settling deeper into the sand and only eight men remained aboard to stand guard. Now just the tip of the propeller was visible. Salvage operations continued until 28 March, by which time 1500 tonnes of cargo and 800 tonnes of fittings had been taken from the *Port Elliott,* which was then left to break up.

The magisterial inquiry found that the stranding had been caused by poor visibility, a set (current) inshore of which the master was unaware and the third officer mistaking a bush fire for the East Cape light. It returned all officers' certificates.

The *Port Bowen*

The last big UK trade ship lost in New Zealand waters — the *Port Bowen* — was also Whanganui's biggest wreck. The ship was far too big to enter the river port, so like all such freighters, planned to anchorage in the roadstead to load the 20,000 sheep carcasses it was scheduled to take to Britain.[17]

Shortly after midnight on 19 July 1939 the ship was approaching the anchorage when it struck the beach a couple of kilometres north of the harbour entrance. Captain Francis Bailey had come in too fast after picking up the Castlecliff light and had missed the leading lights of the anchorage. He had also paid insufficient attention to the echo sounder, and was later blamed for not stopping the ship when a sounding of 80 feet was recorded, indicating shallow water.

The ship had hit a sandy beach, so initially hopes of salvage looked promising. After the local tug was joined by tugs from Wellington and Lyttelton, several attempts were made to pull the ship off, but each ended in frustration as the seas worked the *Port Bowen* further up the beach. A few days later the big tugs departed, and work started on lightening the ship, although winter storms slowed the work and pushed the ship further up the beach. A ramp was constructed to get trucks alongside, but at 40 carcasses a truck, the ship's cargo of 75,000 carcasses diminished only slowly.

The last meat was removed at the end of September when the tugs returned, to fail again, after which the ship was abandoned to the underwriters. Although the Whanganui Harbour Board unrealistically ordered the owners to remove the wreck within six months, its demolition continued until August 1943, when only a little of the lower hull was left to rust away under the sand. Some of the steel recovered from the *Port Bowen* was used to patch damaged ships; some was used to build an anti-tank fence along Wellington's Lyall Bay beachfront.

RIGHT, TOP & CENTRE When the big overseas freighter *Port Bowen* went ashore two kilometres north of Whanganui, a major salvage operation sprang into action. The tugs *Toia* and *Terawhiti* left Wellington to assist the Whanganui tug *Kahanui* and lighter *Morning Light*. When they failed to pull off the ship, the authorities decided to discharge the cargo at low water, as illustrated here. About 3000 bales of wool were discharged in this way.
Top: Gift of Wellington Museums Trust, New Zealand Maritime Museum (2012.0.1297). Centre: photograph by Tesla Studios. Gift of Wellington Museums Trust, New Zealand Maritime Museum (2012.0.6144)

RIGHT, BELOW Ironically, salvage of the cargo became easier as the salvage of the ship became less likely. Eventually the salvors, Wm. Cable & Sons of Wellington, under government contract, ran a temporary pier out to the *Port Bowen* to make it easier to strip the vessel of fittings and then to dismantle the hull and superstructure for scrap. Some of the fittings were used in the wartime minesweeper programme; other material was used by the Wellington hospital and the Whanganui freezing works. Over 6000 tonnes of steel was taken ashore along a special rail track along the beach.
Photograph by C.F. Newham & Company. Gift of J. Grainger, New Zealand Maritime Museum (2015.100.67)

12

Floating Jonahs and hoodoo ships — unlucky ships

This chapter follows the misadventures of four twentieth-century freighters. The first pair, the steamer *Wairau* and the scow *Echo,* traded to Blenheim for many years. One is gone now, but the *Echo* survived many accidents and until recently clinged precariously to existence on the Picton foreshore. The third ship in this floating rogues' gallery, a veritable floating disaster sank off Banks Peninsula over 40 years ago. The last, a trans-Tasman container vessel, foundered early this century after a short career.

The *Wairau* and *Echo*

The *Wairau,* or 'the Jonah Ship' as sailors (a notoriously superstitious lot) called it, began life in 1900 as the schooner *Ronga* for Kerr and Brownlee of Lyttelton. A product of T.M. Lane's famous Totara North yard, the *Ronga* partnered the schooner *Falcon* in the timber trade between Lyttelton and Pelorus Sound. Faithfully built of selected timber (mostly kauri) and very heavily sparred, the *Ronga* was designed for speed. Unfortunately, its tender lines and those heavy spars proved its undoing.

Timber was a one-way trade, so the *Ronga* left Lyttelton in ballast, which the crew dumped overboard as they came up the sound to load. This posed a risk if unexpected gusts caught the *Ronga* empty under sail, and suspicions about the vessel's seaworthiness were roused when it capsized in Pelorus Sound in September 1901 in a squall. A strong nor'easter was blowing, and second mate J.C. Ipsen protested when Captain Ned Petersen ordered the ballast jettisoned. Petersen had a reputation for 'hard driving', and off Fathom Bay, not far from the sawmill, a strong gust sent the *Ronga* over. Luckily, everyone made it into the lifeboat.

In those pre-radio days coasters released pigeons at the entrance to the sound to advise of their arrival, so when the *Ronga* failed to show up, a search began. The vessel was found lying on

OPPOSITE The *Echo* — pictured here at Blenheim wharf — was a typical New Zealand scow, carrying its cargo stacked on deck. The ship has had more lives than a cat, having encountered numerous strandings; collisions; at least one demasting; and most famously sinking after hitting Pencarrow Head. Its exploits even formed the basis of the 1960 comedy film *The Wackiest Ship in the Army,* where it starred alongside Jack Lemmon and Ricky Nelson.
Ref: 1/2-045784-G, Alexander Turnbull Library, Wellington, NZ

The hazards small traders faced at small river ports such as Blenheim were many. Most resulted in minor incidents, but not so the *Neptune*, illustrated here stranded on the Wairau bar in February 1897. It was a total loss.
Gift of Wellington Museums Trust, New Zealand Maritime Museum (2012.0.3963)

its side in Four Fathom Bay. The crew had spent the night in a nearby hayloft and were already stripping gear from the basically undamaged hull.

Once was bad, but surely not twice? Tongues really wagged when the *Ronga* repeated the incident in Pelorus Sound almost exactly a year later. The ship was off the jetty when, 'without warning we were hit. The *Ronga* turned over and the next instant we were floundering in the water,' a crewman recalled. No one was hurt and the lighthouse tender *Tutanekai* helped right and pump out the *Ronga*, but the Marine Department ordered the masts shortened, as it considered the *Ronga* 'overhatted' — too heavily sparred.

By now the *Ronga*'s reputation made it hard to attract crews. Peterson had another close shave in Pelorus Sound at the end of 1905, and on 21 April 1906 the *Ronga* capsized for the third time, this time with fatal consequences. In a hurry to get out of Lyttelton to take advantage of the wind, Petersen had not properly trimmed the 30 tonnes of coal aboard. Since there were no shifting boards to keep it in place, it became a menace to the *Ronga* once things got rough.

Petersen passed the breakwater at 1100 hours under full sail. Later that afternoon a terrific gale sprang up, and the steamers *Talune* and *Pateena* — themselves battling heavy seas — reported seeing the upturned hull of a derelict schooner off Cape Campbell. Several days later the steamer *Gertie* found it off Kapiti Island and towed it in. Fishermen righted the hull and identified it as the *Ronga*. There was no trace of the six crewmen.

In 1908 it re-emerged from a Wellington shipyard as the 143-ton steamer *Wairau*. At first the new name seemed to bring no better luck. On 25 February 1910 it was wrecked at Whanganui in a gale; no one was killed. More extensive repairs followed. The *Wairau* had been back in service for just a few weeks when on 16 May 1911 it hit the steamer *Himitangi* in the Manawatu River and sank. No lives were lost, but it was back to the shipyard for more repairs.

After that, however, things settled down. Under the Eckford flag, the *Wairau* partnered the scow *Echo* and suffering no more than the usual run of minor strandings and machinery breakdowns. The old ship was laid up during the Great

Depression but returned to service in 1942 after the Americans requisitioned the *Echo*. Converted to diesel after the war, the *Wairau* continued to cross Cook Strait until 1951 when it was hulked. Taken to Motueka, it was burned there in 1986 while lying derelict.

Although they have their counterparts in several parts of the world, most notably around San Francisco, scows were the only major trading vessel built especially for New Zealand conditions. Flat-bottomed for beaching and for trading up barely navigable rivers, they were perfect for pioneering conditions. The first, the *Lake Erie*, appeared in 1873. It would be followed by a couple of hundred similar vessels before steam and oil propulsion finally took over from the 1920s.

Scows were tough. Most eventually succumbed to accidents or to old age, but only after surviving accidents that would have finished a conventional schooner or ketch. Scowman Ted Ashby recalled that in his short time aboard the *Rimu* 'she stranded twice, but so versatile were the scows that though strained and twisted out of shape they would soon be patched up and back to work. In my time the *Ngahau* rolled over three times, the *Lena* twice, the second time drowning three people, the *Vesper* once, drowning all hands and the *Glenae* once.' They were almost indestructible, and when things went wrong were easy to salvage — usually by filling them with empty oil drums to provide buoyancy, pumping them out and then towing them back to the shipyard for repairs.

The scow with more lives than a cat was one of the largest, the 126-ton hold-scow *Echo*. This ship — whose exploits formed the basis for the early 1960s comedy film *The Wackiest Ship In the Army* — was built at the Kaipara yard of W. Brown & Son in 1905. Unusually, the *Echo* had an oil engine fitted to assist berthing, however the rest of the time the economy-minded owner expected the crew to use wind power. The *Echo* traded for the shipbuilder the Karamea Steamship Co. and then for Richardson & Co. of Napier, which in 1920 sold the ship to Charles Eckford of Blenheim, for whom it served for 45 adventurous years.

Eckford had entered the Cook Strait trade in 1880 when Thomas Eckford bought the steamer *Mohaka*. By 1920 his son Charles was looking for a small, shallow-draught ship to trade between Blenheim and Wellington. Blenheim? Until the 1960s Blenheim merchants could bypass Picton by using the Wairau River. It was shallow, and more twisted than a politician's tongue, but the river let them unload their produce direct without having to transship goods. The direct service was also popular with farmers supplying the capital.

The Wairau bar was the scene of most of the *Echo*'s mishaps. Fortunately, most were minor. The depth of the bar varied with the weather but averaged 1.8–2.0 metres. Since the *Echo* usually drew 1.7 metres forward and 1.8 metres aft when loaded, there was little margin for error. 'It was often a matter of bouncing across the bar,' Gavin Dobie recalled. 'On most crossings it was usual to hear the bottom of the ship scraping the gravel on the bar.'[1] The *Echo* usually hit the bar with its stern, relying on momentum and the following sea to get across. Naturally, this bent many propellers over the years, and caused some anguish in the Marine Department over unreported incidents.[2]

There were times when the *Echo* ploughed its way across, sticking fast to the bar or stranding on the nearby beach. When that happened, the crew lightened ship and prayed that the weather would hold. Fortunately, it usually did, but during the *Echo*'s worst stranding it spent a fortnight on the bar, bringing Blenheim's direct trade to a halt.

On 17 October 1960 the *Evening Post* reported one of the vessel's more noteworthy clashes with the bar:

> A sojourn on the beach until about the end of this week seems to be ahead of the scow *Echo,* which grounded on the Wairau Bar while outward bound from Blenheim to Wellington on Friday. Her stern grounded on the landward side of the river channel, which runs almost parallel with the beach,

with a boulder bank on the seaward side.

It was unfortunate that the mishap came at a time of weakening tides and that a heavy westerly swell set in which pushed the *Echo* about 20 ft [6 metres] up the beach, for this combination of conditions put her in a position where nothing can be done to move her until the tides improve.

Today the scow was broadside on to the beach with her bow pointing south-eastwards up the river channel and her starboard side high and dry. A track was cut through the shoulder of the beach with a bulldozer to enable motor-trucks to be taken alongside to receive the cargo which was being unloaded — the bulldozer standing by to assist them up the beach when full.[3]

Three days later the *Echo* was on the move. By now empty, it was moved 35 metres towards the river channel by two good pulls on a 'deadman' positioned across the channel. Next morning it was alongside the Blenheim wharf reloading the cargo.

Strandings were relatively mundane events, since once inside the bar the vessel was safe from the destructive force of the waves. So for added spice, the *Echo* added collisions to its repertoire; it hit five other vessels while running in and out of Wellington. Two were Eastbourne ferries, the *Muritai* (30 April 1932) and *Cobar* (9 December 1932). Other victims were the coaster *Gael* (17 January 1935), the launch *Bon* (27 May 1946) and the tug *Toia* (2 January 1931).

Dismastings were also common. Early in the 1920s the *Echo* lost its foremast while beating up Wellington Harbour. A few years later it was the turn of the mainmast, on 31 October 1926 while the ship heading to Blenheim with a full cargo. At 0720 hours, on the last port tack midway between Ward Island and Somes Island and facing a hard-northerly gale, both engines failed, putting a sudden, unexpected weight on to the canvas. The mainmast snapped almost a metre above the aft deckhouse and fell over the starboard side, narrowly missing the men working topside. After Captain Radford stopped the engine, the crew, using the *Echo*'s petrol winch, retrieved the broken mast and the *Toia* towed the scow back to port, where it underwent yet another repair job.

The *Echo*'s worst moment came on 26 November 1932 when it hit Pencarrow Head. A heavy squall struck as Captain William Jarman entered port, leaving him steering blind and unable to change direction, since the *Echo* was not powerful enough to turn around. Just before midnight it struck a rock near the lighthouse, startling the seaman who awoke to the crash of a great jagged rock thrusting through the planking by his bunk.[4] The ten crewmen abandoned ship and made for Scorching Bay, where they landed, some shivering in only singlets and trousers, at about 0130 hours.[5] Meanwhile, the *Echo* — capsized but still afloat — headed off in the opposite direction. Next day searchers found it semi-submerged off Somes Island. The Marine Department accepted that the weather was to blame for the accident.[6]

Requisitioned by the Americans to supply Pacific coastwatcher stations, during World War II the old *Echo,* sporting a couple of Oerlikons, nosed around in dangerous waters, overflown by Japanese aircraft on several occasions.

In 1965 old age and competition from the new rail ferries finally put the *Echo* out of business. On 20 August, sailing to Blenheim for the last time, it went aground just inside the Wairau River mouth soon after midnight. Just minutes before, the old ship had acquired a very special 'cargo' — a launch-load of well-wishers determined to accompany it up the river for the last time. Undaunted, they decided to toast the *Echo* from its position on the mud bank. Blenheim residents who went down to the wharf were kept up with the play by a notice chalked on a crate outside the wharf shed: '*Echo* on the mudflats, due to arrive in Blenheim approximately 4.00 pm today'.

THE *ECHO* — END OF AN ERA

After a brief period 'in lay-up', the *Echo* was bought by J. Gisby of Stewart Island for use as a mother ship off the Chatham Islands. After returning to Lyttelton, it passed into the hands of R. Mason of Picton who retained the old ship until the following July when the Echo Preservation Society bought it, minus the engines.

The society returned the scow to Blenheim for restoration as a public relations office and maritime museum. Sadly, the society's means did not match its ambitions, and the restoration programme failed to get much beyond the planning stage. By July 1968 when the *Marlborough Express* highlighted the *Echo*'s's plight, vandals had stolen many of its fittings and Captain T.S. Eckford was insisting that sinking at sea would have been a more dignified end for it. Stripped of its fittings, the *Echo* eventually filled with water during a flood and sank in the Opawa River. The volunteer fire brigade became a regular visitor with its pumps and appliances.

That should have been the end, but enthusiasts from the Marlborough Cruising Club pumped the *Echo* dry and in June 1972, with the help of several jet boats, towed it down river, over the Wairau bar for the last time and along the coast to the Picton foreshore where, after a very basic 'restoration', the *Echo* served the club as a clubhouse, tourist attraction and relic of earlier, more adventurous times, for 20 years. In 1992 the club sold it to Tim Dare, who re-opened the ship as the 'Echo Gallery' in November 1993 ('Echo Museum' from 2002), serving as a small scow museum and cafe. In 2006, its centennial year, the *Echo* was still up on the hard and up for sale.

Things were looking up when in 2012 she had a new owner, but by 2014 liquor licensing problems were bringing the viability of the *Echo*'s new life into question. That year the vessel was sold to Port Marlborough. Soon after, engineering reports identified the extent of rot in the old scow's timbers, with resource consent being given for its eventual demolition in 2015.

The *Echo* photographed in December 2006, its 100th year, when the old ship was eking out a precarious living as a cafe, bar and museum.
Gavin McLean Collection

The *Turihaua* after modifications to its cargo-handling gear, at Wellington in Richardson & Co. colours. The modifications brought greater efficiency, but no better luck.
Gavin McLean Collection

The terrible *Turihaua*

Another 'hoodoo ship' was the *Turihaua,* the pride — and, for that matter, the sole example — of the Gisborne Sheepfarmers' Frozen Meat & Mercantile Company Ltd's fleet. Built at Bergen in 1948 as the *Anne* and renamed *Sunny Girl* two years later, it was acquired in 1951 to replace the ageing *Margaret W*. With the help of the Union Company (dragged into this deal to protect the interests of its Napier subsidiary, Richardson & Co.), the *Sunny Girl* was purchased for £84,000, spruced up and sent to New Zealand with a cargo of newsprint, steel, prefabricated houses, boilers and tinned foods. It reached Auckland on 24 February 1952.

Four days later at Gisborne, it became the *Turihaua,* a local place name. Comments from a seaman after its loss 11 years later suggested that there might have been something in the name.

The *Taranaki Herald* said that the Maori name *Turihaua*'s interpretation '"is crippled or battered by the wind and water" and this ship has lived up to her name'.

Six days before Christmas 1952, while rounding Tuahine Point in a heavy swell, the *Turihaua* hit a submerged object and started flooding, threatening the engine room. The captain turned around and just made Gisborne before all power was lost. The fire brigade pumped out the *Turihaua* and repairs kept it out of service until June 1953. While in dock, the foremast was relocated amidships and an extra pair of derricks was added to speed up cargo handling.

Six months later on 20 November 1953 the little ship struck the bottom while swinging off Tokomaru Bay Wharf, although the master failed to report the incident. After the damage was discovered the following year during a routine docking, a formal court of inquiry censured him. But when the *Turihaua* was docked again in October 1954, more minor hull damage was found, the result of an unreported incident on 20 May.

That was the last straw for Gisborne Sheepfarmers. In August 1955 the Union Company transferred the ship's mortgage to Richardson & Co., who on 1 September chartered it to Union for the Auckland–Wellington service. But any hope that the jinx had been lifted was dashed just 16 days later when the *Turihaua* ran aground on the western side of Great Mercury Island and became pinned to a reef, bow-up. The master's attempts to reverse off by using the engines failed, and by the early hours of the morning flooding and a falling tide were causing the *Turihaua* to bump. Fortunately, one last blast with the engines freed the ship, and the *Turihaua* limped out to deeper water to wait for Richardson's *Kuaka* to escort it to Auckland, where it endured another round of repairs and courts of inquiry. The court admonished the master and second officer. Repairs (£19,500) were completed by 4 January 1956, by which time the *Turihaua* had become an insurer's nightmare.

Possibly hoping that 'out of sight, out of mind' held true, Richardsons shuffled the *Turihaua* to Nelson for a one-year bareboat charter to another Union Company subsidiary, the Anchor Shipping & Foundry Company. On 8 May 1956, while sailing from Nelson to Wellington, the little ship with the big nose for trouble ploughed into Walker's Rock in a heavy squall. Walker's Rock is an evil place, where vicious tides and swift currents have claimed several ships, and the master of the *Turihaua* must have felt relieved to have escaped with just bumps and dents. The ship re-entered service a few weeks later, and the court of inquiry exonerated the master, to the surprise

> **Walker's Rock is an evil place, where vicious tides and swift currents have claimed several ships, and the master of the *Turihaua* must have felt relieved to have escaped with just bumps and dents.**

of the Marine Department's nautical adviser.[7]

With the exception of a minor stranding at Nelson on 25 October, the rest of the *Turihaua*'s time with Anchor was uneventful. The ship returned to the Richardson fleet in 1 January 1957 to again be chartered to the Union Company. Due to take up a fortnightly run between Auckland and Wellington via the East Cape ports, the day before commencing this the *Turihaua* was leaving the capital for Auckland on 7 February when its engines failed off Steeple Rock. Fortunately the pilot boat *Arahina* was nearby to take it in tow.

Nineteen fifty-seven brought one annoying incident after another. On 14 February the steering gear failed. Fourteen days later a main engine broke down off Gisborne. On 15 March a generator failed. On 11 May the starboard main engine broke down at Portland, obliging the *Turihaua* to limp to Auckland on one engine. The Marine Department had to give permission to sail on this one engine for the next three weeks but restricted the ship to trading between Auckland and Gisborne. On 14 June another steering gear failure sent it crashing into Gisborne's wharf, splintering stringers. On 7 October it had to be assisted into that port after an engine broke down.

The *Turihaua*, now trading as the *Holmbank*, some 14 hours after running aground on Whale Rock near the entrance to Piraki Bay on the south coast of Banks Peninsula. Daylight allowed for an underwater examination, which found that the vessel had been holed in four places and also had a gash along the starboard side. Her back also clearly broken, she was beyond salvage. Nonetheless, as the broken ship creaked and groaned items were hastily retrieved — the ship's wheel, compass, chronometer, and life buoys — before the *Holmbank* eventually broke in two and slid off the rock.

For some time following the sinking the beaches were littered with cargo, which included instant coffee, Disprin, Cadbury Crunchy Bars, and wooden casks of sherry — everything, as one individual noted, that the local farming community needed to both make, and relieve, a good headache!

Jan Shuttleworth Collection

Richardsons had had enough, cursing the Union Company for handicapping it with the *Turihaua* and overvaluing the ship. In December 1957 the agreement was cancelled and the *Turihaua* was laid up at Auckland, its inefficient and unreliable engines (made worse by New Zealand engineers' inability to maintain them) keeping it stuck in port where it spent the next two years, its price, reputation and low speed deterring buyers. Even tied up, the *Turihaua* attracted trouble. In May 1959 the freighter *Indian Reefer* grazed its stern while shifting berths, bending a few stanchions and guard rails. More seriously, in March 1960, engineers — removing the old engine — accidentally flooded the hold bilges.

This re-engining was a final gamble. For £48,000 (split between Union and Richardsons) the twin A/S/ Bergens Mek Verk engines were replaced by a more familiar Crossley V-8. Trials took place in July 1961 and, after one breakdown, the *Turihaua* re-entered service, under charter to the Union Company's Wellington affiliate, the Holm Shipping Company, from 12 January

1962. Shortly afterwards Holm bought the ship, which it renamed *Holmbank* for its trade between Onehunga and South Island ports.

New name, same old bad luck. The *Holmbank* lost money during its first year and ran aground in its second — this time for keeps. On the evening of 20 September 1963, while on a voyage from Timaru to Wellington with over 400 tonnes of general cargo, the *Holmbank* struck a rock in Peraki Bay on Banks Peninsula. Captain Walter Home had set a course to pass 1.75 miles clear of Akaroa light. Weather reports of low visibility in the area of the light meant that the crew did not worry too much when the light did not appear as expected. But unknown to them, a strong current had been pushing the ship off course, towards the land, and it went aground at 2022 at high tide.

By next morning the tanker *Tanea* and the freighter *Waimea* were standing by while a tug steamed from Lyttelton. The sea was calm and the wind slight, but by the time the tug *Lyttelton II* reached the *Holmbank* its hull was two-thirds full of water. Any attempt to pull it off would only have resulted in its sinking. An engineering firm assembled salvage equipment, but next morning the ship broke its back with a loud cracking sound. Rising seas added to the damage, and around dawn on the 22nd the *Holmbank* split in two and slid off the reef, the bow sinking first.

The inquiry blamed an unanticipated set and the crew's complacency due to that weather report of low visibility off Akaroa. It was an unexpectedly uncontroversial ending to the story of the unluckiest ship on the coast.

The *Sydney Express*

Twenty-three years ago a German shipyard built two small container ships. They were identical. Both served the same line and ran on the same trans-Tasman trade. One caused its owners very few problems and still serves other owners. The

The *Sydney Express*, photographed in happier times on Otago Harbour.
Ian Farquhar

The Wellington wharf police towing the upturned *Maria Luisa* back to shore after it had collided with the container ship *Sydney Express*.
Ref: EP/1996/3725/3A-F, Alexander Turnbull Library, Wellington, NZ

other was a floating Jonah and no longer exists. The ship was the *Sydney Express*.

The *Sydney Express* was built for charter to the Tasman Express Line (TEL) — a joint venture between New Zealand companies Refrigerated Freight Lines, Scales Corporation, McKay Shipping and the Australian shipping agency Hetherington Kingsbury Ltd. In 1985 TEL chartered two small cellular container ships, the *Auckland* and *Canterbury Express*. For years the company dithered over replacement vessels before ordering two new ships to be time-chartered from a German firm. The *Wellington Express* and *Sydney Express* entered service in 1996.

Although they were lightly built, vibrated at speed and were noisy, their Achilles heels were their MAN B&W medium-speed diesel engines — not previously used in ocean-going ships. Between 1996 and 1999 there were several embarrassing breakdowns. But that was nothing in comparison with what occurred just inside the entrance to Wellington Harbour late on the night of Sunday 29 December 1996.

Captain Rod Lott was on the bridge with third officer Craig Lucena and helmsman Stephan Taylor as his ship headed out to sea. The weather was fine and warm. The regulations require ships of more than 500 GRT to keep as close to 'the outer limit of the channel or fairway which lies on her starboard side' as is practicable. Smaller craft (under 20 metres) were required not to impede the navigation of a vessel that could navigate safely only within a narrow channel. In practice, big ships, while keeping to their correct side of the channel, also stayed close to the centre of that channel to maximise sea room and manoeuvrability.

In 1996 small fishing boats and pleasure craft were encouraged — but not required — to report their movements by radio. The inward-bound trawler *Maria Luisa* reported to the Beacon Hill station when two miles off the buoy, but the *Sydney Express* did not hear this because the port's 'open-loop' radio system did not oblige Beacon Hill to pass on the information. Ships were merely expected to monitor VHF channel 14. A minor player in the tragedy, a yacht, was on channel 16.[8] The *Sydney Express* bridge crew, busy with other things, turned down the VHF volume, and the trawler skipper, whose radar was not working, told Beacon Hill that he was about to switch off. This was unfortunate. At 2223 a conversation between Beacon Hill and the ferry *Aratika* mentioned an inbound fishing boat. After the *Maria Luisa*'s skipper said that he was about to stop listening, another conversation between the *Aratika* and the station said that the *Sydney Express* was 'between the leads'.[9] Both vessels missed valuable warnings.

As the *Sydney Express* tracked towards Falcon Shoals Beacon, between Karaka Bay and Ward Island, at about 12 knots, Lott and Lucena saw two vessels on their starboard bow, out near the harbour entrance. They thought that each was showing a red (port, or left side) sidelight only and mistook the yacht *Soundsgood* and the *Maria Luisa* for yachts under sail.[10] Lott estimated that the vessels were on the port side of the channel and crossing over to the starboard side. Some time about then the light illuminating the forecastle was switched off as the ship worked up to about 14–15 knots.

The first close encounter was with the 11-metre sloop *Soundsgood* (*Sounds Good,* in some reports)

As the vessels came together, Lott gave one medium blast on the whistle 'to wake him [the trawler] up' and moved the propeller pitch control lever to full astern. But it was too late.

which was motoring in at 4.5 knots, exhibiting a single tricolour lantern; the white light it should have been displaying as a power vessel was not on. The yachtie estimated that the ships passed in the channel near the north end of Barrett Reef about 50–100 metres apart. At 2230 Lucena rang for full sea speed and walked to the port bridge wing to take a visual bearing of Pencarrow light.

Lott had ordered helmsman Taylor 'starboard five' to increase the passing distance between the ship and the yacht. By then he was watching out for the other vessel 'fine on his starboard bow and still showing a red'.[11] Worried that the *Sydney Express* was now too close to Barrett Reef, Lott decided to 'continue showing the *Maria Luisa* a green' (green sidelights are on the starboard or right side of a ship) and alter course to port. Believing himself to be closer to Barrett Reef than investigators later thought, he felt that he had insufficient sea room to alter course further to starboard, and he knew from past experience that small vessels 'usually nipped around to port and showed me a green'.[12]

From this point Lott and Taylor disagreed on what happened. Lott told TAIC that after passing the yacht 'he ordered the helm to midships and possibly steady on a course which he vaguely recalled being 200 degrees; after a short delay he gave a new course to steer of 185 degrees (10 degrees to port of the 196 degrees course the helmsman had last steadied on)'.[13] Lott said that he felt the ship heel slightly to starboard, looked at the rudder angle indicator and saw it reading 15–20 degrees to starboard, shouted out 'bloody hell, you're going the wrong way man!' and ordered the rudder hard to port.[14] Lott told the MSA that by ordering 'hard a port', he hoped to avoid coming too close to Barrett Reef and 'to enable him to continue to show a green light to *Maria Luisa* so that the fishing vessels would be discouraged from crossing ahead and encouraged to pass clear down the starboard side of *Sydney Express*'.[15]

Taylor testified that he did not hear any order to amidships the helm or steady on a course. He said that the next order he received after 'starboard five' was 'steer 185'. He was still determining which direction 185 was and was about to apply port helm when Lott told him he still had the wrong helm on. He denied applying 20 degrees of starboard helm and that he was still applying five degrees of starboard helm in accordance with the last order he had heard from Lott. As the vessels came together, Lott gave one medium blast on the whistle 'to wake him [the trawler] up' and moved the propeller pitch control lever to full astern. But it was too late.

What happened in the trawler's wheelhouse in the final minutes is anyone's guess. The only survivor, Brendan Gee (an off-duty trainee), recalled skipper Mel Webster say, 'Oh look, there's a big ship. What the hell is it doing?' before applying starboard helm. Gee looked up and saw the red and green lights of the ship heading right for him.[16] At about 2233 hours the *Sydney Express*' bulbous bow struck the trawler's port quarter.

While Lucena warned Beacon Hill, Lott put the ship into the middle of the channel and turned around to launch a boat. Brothers Mel (skipper) and Pat Webster, their brother in-law Paul Sundgren, cousin Andrew Bettison and friend Robert James Smith had drowned, trapped inside. Gee was winched aboard the ship.

The principal reports by the Transport Accident Investigation Commission (TAIC) and by the Maritime Safety Authority (MSA) were released in August 1997. Both found that the crews of the yacht, the trawler and the container ship were in error. TAIC found that the *Soundsgood* was not exhibiting the correct navigation lights and had impeded the passage of the *Sydney Express*. The ship's deviation to avoid the yacht placed that ship further to starboard in the channel than normal. TAIC found that by proceeding up the port side of the channel the *Maria Luisa* impeded the passage of the *Sydney Express*, thereby contravening the collision regulations.[17] Unfortunately, the container ship's crew did not make proper use of the radar equipment fitted or sound a warning with the ship's whistle. Captain Lott's decision to alter course to port was based on inadequate information as to the relative positions of the vessels and was made at a time 'when his workload was high and his situational awareness was low'.[18] Poor bridge resource management on the *Sydney Express* caused the master's high workload and the low situational awareness among the bridge team. Altering course to port was not the best action to aid avoiding collision. The *Sydney Express* had adequate sea room to move further to starboard to avoid the *Maria Luisa*.[19] In putting the helm the wrong way the helmsman added to the master's workload but did not cause the collision. The master's unclear helm instructions under pressure, the poor ergonomics of the helm, and the helmsman's inexperience, all contributed to the wrong helm application. TAIC also found that both masters missed vital information by not closely monitoring VHF channel 14.[20] It recommended giving ship crew Bridge Resource Management (BRM) courses and fitting an Automatic Radar Plotting Aid (ARPA) to reduce the workload on the bridge team.

The *Sydney Express* suffered only minor dents

The *Maria Luisa* at Nelson for repairs.
Michael Pryce

and scratches, but the 19.8-metre *Maria Luisa* was forced under the *Sydney Express* and up again, sinking in shallow water off Seatoun. Lifted by the floating crane *Hikitia*, the *Maria Luisa* was towed to Nelson for permanent repairs. The survivors' families — dissatisfied with the official reports — maintained that the *Sydney Express* was further south than claimed and clear of the narrow channel. In 2003 the MSA appointed an independent examiner to assess new evidence, but while disagreeing with minor aspects of the earlier reports, he blamed the collision on a long and tragic chain of errors and found that 'the *Maria Luisa* was the vessel that bore most of the blame'.[21]

13

Coasters in crisis

By the mid twentieth century it was rare for freighters to sink at sea. Radio, aircraft and radar kept a watch on the sea lanes, removing much of the uncertainty. But not all of it. In the course of just a decade three relatively modern motor vessels — the *Holmglen, Maranui* and *Kaitawa* — sank off the coast, all with heavy loss of life, and one, the *Calm,* was lucky to be salvaged.

Calm before the storm

On the afternoon of 13 July 1956, Captain C.M. Davies, the master of the coaster *Calm,* was praying for weather that could live up to the name of his charge. The 787-ton freighter, just six years old and the pride of the Canterbury Steam Shipping Company's fleet, had cleared New Plymouth and was butting through a southerly gale bound for Dunedin with a cargo of pork and general goods.[1]

By the time he was off Opunake, Davies, a man with 38 years' seagoing experience to his credit, was worried. Very worried. Mountainous seas had flung the ship off its course and were causing it to lose steerage way. Although it was in no immediate danger of foundering, the time had come to seek shelter. Later that evening Davies turned around and made for New Plymouth.

With the sea behind it, the ship rode much easier. Then, at 0118 hours on 14 July, in complete darkness and without any warning, the *Calm* crashed onto the rocks off Waiweranui Point, Taranaki, and started taking in water. An eerie darkness enveloped everything, including the Cape Egmont light, which had been visible on the way south but which did not reappear until much later, six miles to the south. Davies ordered the crew to ready the lifeboats but, believing it too dangerous to leave the vessel in the dark, withheld the order to abandon ship.

Daybreak revealed the beach just 100 metres away. By that stage of the tide the *Calm*'s bows were far enough out of the water to enable the 14-man crew to step ashore safely and raise the alarm at a nearby farm. The farmers, the Soles, gave them breakfast, and Lloyd Sole used his tractor to transport the men's personal effects

OPPOSITE Well wrapped up against the winter winds, company officials supervise the lightening of the *Calm*'s cargo. This photograph, taken at low tide, shows how far up the beach the *Calm* had gone and also indicates why salvaging it was so difficult.
PHO2010-0711, Puke Ariki

ashore. Even the ship's cat, chaperoned by Paddy Leahy, reached the beach with dry paws!

By the time the company's marine superintendent, Captain W.D. Drake reached Waiweranui Point later that afternoon, everyone knew they faced an uphill battle. The *Calm* was wedged firmly on a reef of boulders jutting out from the point, about 65 metres from the shore. Heavy seas were pounding its starboard side and it was taking in water. With a south-east gale blowing and heavy rain falling, things looked bad.

By Sunday morning the seas had calmed sufficiently for Drake's men to board and inspect the ship. The port and starboard tanks were leaking oil, and the double bottom freshwater tank was full of saltwater; the ship was badly damaged, but fortunately the hold and engine room were dry.

Drake's men rigged up a flying fox and started lightening the ship, seven sides of frozen pork at a time. That afternoon they were joined by additional workers and a bulldozer. Weather conditions were appalling in the cold southerly winds. Much of the cargo and stores had been removed by the afternoon of the 17th, when the Union Company's tug *Taioma* arrived. It was joined by a small powered punt which quickly fouled the *Calm*'s lines and went ashore. In a matter of minutes, the rescuers found themselves in need of rescue. Only quick thinking by the bulldozer crew averted tragedy and saved the lives of two men working on the punt.

Weather conditions prevented work the next day, and only the promise of a higher than normal tide the following day kept salvage hopes alive. By 1930 hours on the Thursday everything that could be moved off the ship had been taken ashore; the *Calm* was as light as it would ever become and was now bouncing about in the surf. With its engines going full astern and pulling on the wires, it eased its way off the rocks and out into the comparative safety of deep water.

Four hours later the ship was safely alongside New Plymouth's Newton King Wharf, where 500 people greeted it. An equally large crowd, which this time included the Soles (who returned the ship's cat just before sailing time), watched the *Calm* leave again, escorted by the *Breeze*.[2] Permanent repairs, which cost £65,000 and took three months, were carried out at Port Chalmers from 27 July. Inspection revealed that the plate keel and the bottom plating were extensively damaged.

The company recovered compensation of £40,000 from the Marine Department whose Cape Egmont lighthouse had gone out after a power failure just before midnight. Since the *Calm* had no radar or direction-finding equipment, the lighthouse's failure was held to be the primary cause of the stranding.[3] The lighthouse had not been manned since 4 January, mariners had not been notified of this, and the department's notification system to the nearby part-time keeper had been inadequate.

Once repaired, the ship served its owners well- surviving a minor collision with a Port Chalmers fishing boat in May 1963 and a cargo fire off the South Canterbury coast in August 1966. Sold to Asian owners in 1971, the former *Calm* finally sank in the South China Sea on 3 July 1990 under the name *Angkor 2*.

'A rock and roll ship' — the *Holmglen*

On the evening of 24 November 1959 Captain Henry Williams of the tanker *Tanea* was walking with Oamaru harbourmaster Captain John Hancox to the home of Shell Oil's local manager. Six hours earlier the men had watched the coaster *Holmglen* clear the dolphin at the end of Holmes Wharf on its way to Wellington and Whanganui. The *Tanea* had also been scheduled to sail that afternoon, but Williams had delayed his departure because he was concerned about manoeuvring his much larger ship out of port in bad weather. At 2100 hours as the men chatted, 'an exceptionally violent south westerly squall' passed over Oamaru, nearly blowing them off their feet. 'I

The *Holmglen*, which sank with all lives after leaving Oamaru on 24 November 1959. As the radar above the wheelhouse shows, this coaster was well equipped for the time.
Photograph by Cliff Hawkins, Wellington Museums Trust, New Zealand Maritime Museum (2012.0.2393)

expressed to Captain Hancox at the time how wise we were to have cancelled the *Tanea*'s sailing as we could well have been caught on the knuckle of the wharf,' Williams recalled.[4]

By the time they reached the house, the phone was ringing to say that distress signals had been received from the *Holmglen*. At 2112 the Taiaroa Head radio station picked up a message. After giving the ship's position, the voice, believed to be that of Captain Edward 'Joe' Regnaud, advised: 'Heeling hard to port. Accommodation awash. Crew attempting to launch boat.' That was it. Nothing further was heard from the ship. The signalman, R. Malthus, repeatedly called '*Holmglen* — are you receiving me?' but got no answer and asked the Awarua marine radio station to call for help. By then seas were heavy, it was raining and it was obvious that the ship must be in serious trouble.

A small fleet rapidly converged on the South Canterbury coastline, battling heavy seas and deteriorating weather. The master of the *Cape Ortegal* reported that the south-west gale was blowing at force 8, with gusts reaching force 9, and that the seas were very high with a heavy swell. They were too much for the Northern Company's *Tawanui*, but the freighters *Cape Ortegal*, *Korowai* and *Holmburn*, the navy launches *Tarapunga* and *Takapu*, the Oamaru fishing boat *Venture*, two trawlers from Akaroa and about a dozen Timaru boats managed to join in the search. It was dangerous work — one Timaru fisherman feared that the wreckage and debris in the sea might start one of the *Rambler*'s planks.[5] Even the navy found it hard going. The *Takapu*'s commander reported that his craft's rubbing strake was slightly damaged, a ladder on the port side of the bridge was broken and 'two metal vents simply vanished … There was chaos inside the launches.'[6]

> '*Holmglen* was the strongest built and the best equipped ship in New Zealand. Of all the ships in New Zealand likely to come to this end I should have picked the *Holmglen* as the least likely.'

Ironically, the sad duty of finding the wreck fell to another Holm ship, the *Holmburn*. Diverted to investigate a spreading patch of oil and wreckage, it located the wreck by echo-sounder in 55 metres of water. The ship radioed the Timaru harbourmaster: 'Coming across more wreckage all the time. Weather is freshening all the time.' Then the *Holmburn* marked the spot with a red anchor buoy and left the area. Two fishing boats, the *Nella* and the *Craigewan,* picked up two of the three bodies recovered. Shortly afterwards, worsening conditions forced the Timaru harbourmaster to order the remaining boats back to port. The third body was later picked up by the *Norseman,* whose skipper, alerted by the sight of a large flock of seabirds sitting on the water, found the 'near skeletal remains' of third engineer Wilfred Harding.[7]

The *Holmglen* had a crew of 15: Captain E.J. Regnaud, K.D. Billinghurst (chief officer), H.L. Barker (second officer), R.A. McFoster (chief engineer), A.J. Wolgast (second engineer), W.H. Harding (third engineer), S. McKenzie, J. Cleary, D. Whorlow, H. Wetherby and A. Pemberton (able seamen), G.J. Boyce (ordinary seaman), J. Anson (cook), A.S. McClellan (steward) and J. McEwan (wiper).

A stunned John Holm confessed that 'it is a complete mystery. *Holmglen* was the strongest built and the best equipped ship in New Zealand. Of all the ships in New Zealand likely to come to this end I should have picked the *Holmglen* as the least likely.' During the next few days, the other Holm ships dropped wreaths over the site of the wreck while the company waited for the court of inquiry.

No ship is unsinkable, but this one, less than five years old and fitted out to supply sub-Antarctic depots, should not have foundered when it did. The 485-ton coaster had sailed from Oamaru at 1540 on the 29th with a mixed cargo from Dunedin and Oamaru. That day the seas were moderate, and although it was carrying 28 tonnes of deck cargo, the *Holmglen* was above its marks. 'I was impressed with her general appearance, everything being apparently well stowed and lashed in a seamanlike manner,' Captain Williams later testified. 'I accompanied Captain Hancox down the wharf and watched the vessel swing on a head rope round the knuckle at the end of the wharf. She listed slightly to port when the weight came on the rope but recovered quickly to an upright position when the rope was slackened indicating that she was stowed in a stable position.'[8]

Much of the evidence centred on that deck cargo. The *Holmglen* was licensed to carry 50 tons (50.8 tonnes). When it left Oamaru, there was 28 tonnes on deck, lashed to the No. 1 hatch, where the ship's forecastle offered protection from heavy seas. But was it stowed correctly? Intriguingly, two days after the disaster the Oamaru superintendent of mercantile marine reported that the stowage of a tier of oats at the bottom, flour above it, topped off with oats was unusual: 'According to both Captains Hancox and Williams ... this was not the method they would have had this cargo stacked,' he reported. 'They pointed out to me that it only needed one bag of oats to burst through getting wet or the lacing coming away and the whole stack would become unstable. As the stack was secured with a cargo net and tarpaulin this would cause it to hang like a big bag and would naturally go to the leeside.' J.C. Winders also added that 'the engine room door on the *Holmglen* was unlatched when the vessel left Oamaru, this was apparently observed by members of the crew of the "Tanea"'.[9]

At the inquiry, however, witnesses agreed that the cargo was safely stowed. Allan McKay, president of the Oamaru wharfies, recalled that work stopped after the first tier had been stowed. He heard the chief officer tell the foreman that he could pump out a ballast tank. 'I heard the remark that this tank would give 30 tons and I got the order to continue loading.'[10] Eric Tutty, the foreman, confirmed that the chief had pumped out the tank. Before that, 'the vessel appeared to me to be a little short of her loadline mark, but the Chief Officer Mr Billinghurst informed me he had about half an inch to spare'.[11] The deck cargo, 160 sacks of oats and 198 bags of flour, was loaded on a double layer of dunnage and covered with a tarpaulin before being lashed down. Tutty confirmed not seeing any water in the hold. One wharfie thought there was something unusual about that deck cargo. Harry Burnett recalled that it was 'straight side for about six feet when it is usually stepped in about half a bag per tier. My gang was instructed to do this.'[12]

The inquiry reprised horror stories about the ship's performance in heavy seas. A former chief officer related how the *Holmglen*'s accommodation had been awash while running before a storm off Campbell Island. In May 1958, while crossing Cook Strait, a deck cargo of coke had broken loose and blocked the scuppers and wash ports, forcing the deck crew to jettison the sacks on the port side to clear the wash ports and get rid of the water. Had this happened and the *Holmglen* broached?

Rowland Masterman, who had served briefly on the *Holmglen,* said it was very hard to steer. 'With any sea just on the bow, we had to rig the foresail. She was underpowered and would almost stop in anything of a sea. She would ship water quite heavily over the bulwarks, forward of the mast housing.'[13] J.D. Garrick, a wiper and former crewman, called the *Holmglen* 'a rock and roll ship', alleging that in heavy seas the tea pot and knives flew from the mess room to his bunk.[14] Patrick Gordon, a former mate, recalled often taking water out of the forepeak: 'On a lot of occasions we had to remove eighteen inches or two feet of water out of the chain locker ... there was sufficient water down there to put her down by the head.' But despite that incident with the coke bags in Cook Strait, and the fact that 'the *Holmglen* had to be watched in a following sea', Gordon considered the vessel 'well found and well equipped'.[15] So did Alex Grieve, a former *Holmglen* master. The ship had never pooped a sea in his four months aboard; when asked if he had considered the ship unseaworthy, Grieve replied, 'I did not.'[16] Captain 'Tup' Cosmo Keith, Holm's marine superintendent, thought 'she wasn't good [at steering], but she wasn't notoriously bad. She was an average little coaster.'[17]

The physical evidence did nothing to solve the mystery. 'Operation Holmglen', the navy's underwater camera survey, found the ship sitting upright on the seabed:

1. A foresail (staysail) was rigged and extended about half way up the forestay — this had been fitted to provide emergency propulsion if the ship's engines failed while servicing a sub-Antarctic island, but masters sometimes used it to steady the ship under certain sea conditions.
2. Both hatches were undamaged and secure; there was no sign of deck cargo.
3. The starboard weather door leading from C hatch deck along past the master's cabin was wide open but not clipped. All wheelhouse doors and windows were closed, as were all port holes.
4. The port sea boat was secured. The starboard one was missing (it had been recovered, empty, on the surface) but the

boat's davits had not been turned out, indicating that it had not been launched by the crew.
5. The only visible damage was to guard rail stanchions and guard rails on the starboard quarter and aft on the boat deck, which were stove-in and crushed, the engine staff was lying on the deck, and damage to the railing around the ladder leading from the accommodation deck to the boat deck.

The debate continued. Timaru fisherman Johnny Inkster wondered if record cold temperatures during construction had weakened the welding, causing a plate to fail. Former mate Bob Jenkins recalled that after repairs to the mast house watertight doors six months earlier, the handles had been put back incorrectly. Could heavy seas have forced the lever down and let in water? In 1999, however, underwater filming for the *Shipwreck* television series showed that both doors to the mast house were still closed.[18]

The court failed to find the cause for the catastrophe. Its report merely stated the obvious, that 'the cause (of the sinking) arose with great rapidity and the vessel foundered with great rapidity'. It ruled out an engine explosion, the opening of plates, collision or striking a submerged object or rock. The sinking was a hot topic amongst shipping people for years. Most agreed that the *Holmglen,* always considered somewhat 'cranky', had lost way, been overwhelmed by a sudden wind gust, and had sunk when the cargo shifted.

The court made 17 recommendations. The main ones were for a re-examination of the design of small, low freeboard ships, making it compulsory for small ships to carry inflatable life rafts and for the Marine Department to tighten up deck cargo licences.

A few months later the Minister of Marine received a letter from Djakarta. The writer, Captain E.F. Rainbow, had skippered the *Holmglen* and had a theory: 'It was my experience with this vessel that when loaded and running before a sea she was dirty aft and in the habit of pooping seas,' he advised. 'It is my considered opinion that as the *Holmglen* was running before a heavy southerly sea the vessel pooped a sea which flooded the galley and saloon and water poured down the accommodation ladder into the Officers' Quarters. Water would also enter the Crews Quarters from the starboard side.' Before the *Holmglen* could recover, it pooped a second sea and sank stern-first. Noting the evidence that neither boat had been launched, Captain Rainbow argued that if the *Holmglen* had been listing, the crew could have launched the boat on the lee side; if, however, the ship sank stern-first, it would have been impossible to stand upright and untie either boat. Rainbow asked for the inquiry to be re-opened, but his letter was filed.[19]

Death off Pandora Bank — the *Kaitawa*

The next major loss was the Union Company's collier *Kaitawa*. Once again, a well-found ship sank with all hands. The 29 lives lost made it our worst maritime disaster in more than half a century. The 2485-ton *Kaitawa* was an 'AC' class coaster, built by Henry Robb at Leith, Scotland, in 1949. Like most Union Company workhorses, it was slow but sturdy, ideal for the West Coast coal trade.

On 20 May 1966 the *Kaitawa* left Westport for the Portland cement works, near Whangarei. The next day it returned to port and anchored off the bar to transfer the second officer, Richard Oakton, who was ill, and to pick up his replacement, Michael Jenkins. At 1313 hours on 21 May the *Kaitawa* resumed its voyage. Oakton, who had collapsed while the ship was off Karamea, did not know how lucky he was.

As the *Kaitawa* motored north the weather steadily worsened, causing Captain George

Typical of the Union Company's postwar fleet, the *Kaitawa* distributed West Coast coal around the country.
Gift of Wellington Museums Trust, New Zealand Maritime Museum (2012.0.6733)

Sherlock to revise his estimated time of arrival several times. His ship, not known as one of the 'slow greens' for nothing, was crawling in the face of a howling gale. At 1840 hours on the 23rd, when the freighter *Cape Horn* passed the *Kaitawa* about five miles west of the northern tip of the Pandora Bank, about 12 miles from Cape Reinga, the weather was atrocious. There was a heavy sea, thick rain and wind gusts up to 35 knots. The *Cape Horn*'s Captain Thomas Edge reported seeing the masthead lights and red side lights of an unidentified freighter (it was also picked up on radar). At 2000 the same ship was observed on the radar about 20–30 degrees abaft the *Cape Horn*'s port beam, about 4–5 miles astern. The weather was now 'becoming boisterous'.[20]

The British thought nothing more of the ship they had passed but then, 59 minutes later, Auckland radio picked up a PAN (an emergency signal denoting danger but not imminent danger) from the collier. Another minute later it heard the more serious and urgent Mayday message from the ship, indicating a rapid change in fortune.

It read: 'Mayday (repeated nine times) *Kaitawa* ZMVC (repeated three times). Position (word or words missing) … 10 miles Cape Reinga. Bearing zero three five (word or words missing) … require immediate assistance.'[21] This position would have placed the ship about a mile west of Pandora Bank, a shallow patch just south-west of Cape Reinga. Atmospheric conditions brought brief interference from Radio Adelaide, and by the time that had been sorted out, the *Kaitawa* had fallen silent. No more was heard from the ship.

A major search and rescue operation swung into action. The *Cape Horn*, about 15 miles to the south, swung around at 2118 and retraced its course. At 2350 crew spotted a red parachute flare 5–10 miles to the west, very near the Pandora

Bank. The radar, badly cluttered by weather conditions, found nothing, and lookouts saw no more flares. 'At this time the wind was steadily increasing to storm force 12 with heavy breaking seas and high swell,' Captain Edge recalled. 'Our own vessel becoming un-manoeuvrable, therefore decided, in view of lee shore, to heave to.'[22] The *Cape Horn* resumed the search at 0936 the next morning, but found nothing and resumed its voyage at 1600 hours.

By then there were plenty of searchers. The air force sent up Sunderlands and Dakotas, the navy dispatched the training ship *Inverell*, and the Northland Harbour Board sent its ocean-going tug *Parahaki,* but all to no avail. Indeed, the rescue ships struggled to make any headway; the *Parahaki,* normally capable of 15 knots, struggled to get 10. In all, 14 ships, eight aircraft and several hundred shore-based searchers took part.

By mid-afternoon a Sunderland had located an oil slick a mile north of Pandora Bank and the first pieces of wreckage were being picked up from the beaches; the discovery of an interior staircase and panelling from either the smoke room or the captain's cabin found embedded with sand, gravel, coral and shell, suggested that the ship had hit the bottom upside down, since the wreckage had not yet reached the shore.[23]

Search parties found the 11 kilometres of coastline south of Cape Maria Van Diemen littered with pieces of timber and broken fittings. The wife of an assistant lighthouse keeper told the *New Zealand Herald* that 'pulped would be a better way to describe some bits; the wreckage ranges from little pieces the size of a pencil to bits seven or eight feet long'. One party, which included the Cape Reinga schoolmaster, picked up 13 lifejackets; all were in good order. Later two lifeboats were found in Twilight Bay. Another party found part of a lifebuoy with *Kaitawa* stencilled on it. Despite extensive searching by

Coal was no longer king in the 1950s, when this photograph was taken, but the wharves at Greymouth (pictured) and Westport were still busy with colliers such as the *Kaitawa, Konini* and *Kaiapoi.*
Gavin McLean Collection

Sunderlands, no trace was found of any other lifeboats or intact life rafts.

In the absence of evidence, it was assumed that the ship had foundered so quickly that most of the crew had not had time to abandon ship. At least one person had inflated the wrecked life raft and discharged the emergency flare, but he could not have survived for long. Only one body was recovered, that of motorman John Wright.

On 8 June the research ship HMNZS *Tui,* using sophisticated underwater camera gear, located the *Kaitawa* at a point 246° 20', less than five nautical miles from Cape Reinga light. It lay at a depth of about 50 metres, upside down, heeled over to port at an angle of about 160° to the upright. The superstructure had vanished completely, either torn off while the ship was drifting capsized or crushed into the hull when it hit the bottom.

Later that month divers from HMNZS *Inverell,* after examining the *Kaitawa*'s sister ship *Kaitangata* at Auckland, dived on the wreck site. They discovered a hole, about three metres by two metres, in the *Kaitawa*'s hull in the approximate vicinity of her No. 3 hold; an explosion was ruled out and the hole was attributed to a ship's frame being pushed through the hull. They also found that the hatch covers on the Nos 1 and 2 cargo holds had gone. There was no sign of coal in the wreck or nearby, which indicated that the ship might have drifted some distance after striking trouble. Also visible were two depressions, running more than 32 metres along the bottom of the ship's hull, and slight damage to the two propellers. Lieutenant Merrick had a quick look at what remained of the midships structure. 'It was just possible to see just inside what had obviously been living quarters. There was no sign of any bodies but I saw a mattress still wrapped in a sheet.'[24]

Some people thought that this indicated that the *Kaitawa* had struck the Pandora Bank, a shallow patch of water about five miles long and three miles wide. Greymouth harbourmaster Captain H. Gordon felt that Captain Sherlock probably did not know that he had entered the danger zone until it was too late. HMNZS *Kiama* surveyed the bank during the search and discovered that it 'extends approximately one mile north-west of its charted boundary and that the Pandora contained a previously undiscovered "underwater "hill" in the approximate position in which *Kaitawa* would have been at 2100 hours on 23rd May if she were steaming along the normal route from her 2000 hours position'.[25]

There was about 6.5 metres of water over the bank at the time. Although the *Kaitawa* drew 5.8 metres, the 6-metre swell would have had it striking the seabed at just over nine metres. After hitting once or twice, it was surmised that the ship would have broached to and lay beam-on to the seas, lurching forward and causing the coal to shift. The end must have been quick. Any on their feet would have been thrown onto the decks. Those in their cabins would have been trapped and helpless. It would have been impossible to put on a life-jacket as the ship capsized and rolled like a log, breaking off its masts and superstructure before sinking. Anyone who got off the ship would not have survived in the terrible seas.

The court of inquiry heard this and other theories but was forced to conclude that, in the absence of evidence from survivors, no cause could be stated with certainty. It did, however, favour the theory presented by nautical surveyor Captain E. Milroy. Milroy believed that the position given by the Mayday message was wrong, believing the ship to be 7–10 nautical miles from Cape Reinga light. The captain speculated that as the *Kaitawa* wallowed in the trough of the sea, a wave burst in the port forecastle door leading to the crew's accommodation. As the water rushed in, the ship started listing sufficiently for Sherlock to change his PAN message (less urgent than 'Mayday') to Mayday. The more the *Kaitawa* listed, the worse the storm damage became. Wave after wave struck it, splintering the bridgework and other parts of the superstructure. At that point the crewmen, mustering before abandoning ship, were probably swept overboard.

From then the *Kaitawa* was out of control, drifting towards the Pandora Bank, its engine failing at some point. Almost certainly abandoned, it would have drifted past the *Cape Horn* (the red parachute flare probably came from the life raft) and onto the bank where, down by the port bow and low in the water, the ship would have struck the seabed, causing the long indentation later found by the divers. At that point the *Kaitawa* probably capsized, losing its hatch covers, cargo and possibly some of the superstructure. Still floating upside down, it drifted losing its residual buoyancy and plunging to the seabed sometime before dawn.[26]

The inquest, presided over by the Auckland coroner, found that all 29 crew had drowned. Although there were demands for further investigations, divers reported that conditions for diving on the wreck were extremely difficult. A surge on the seabed and jagged pieces of metal protruding from the hull made it dangerous. Lieutenant Merrick ruled out using explosives because they might cause an explosion from the build-up of gases or attract sharks.[27]

The men lost were Captain George Sherlock, chief officer Robert McEwen, second officer Michael Jenkins, radio officer Phillip Mowat, chief engineer Oswald Horrobin, second engineer Garry Emmerson, third engineer James Fox, fourth engineer Royce Williams, electrician William Underwood, boatswain Raymond Hill, ABs Bruce Oliver, Anthony Meekin, Tex Frederick Walker, Gerald Casey, James Wilson, and Victor Clarkson, OSs Kerry Sheldon and Charles Pulekula, deck boy Ian Hayward, crew orderly Thomas Byrne, motormen John Wright, John McLean, James Mcleary and Charles Fletcher, chief steward J. Pickles, assistant stewards Geoffrey Jones and J. O'Connell, chief cook Bert Smith and assistant cook D. Collett.

'Real-life drama enacted off the Coromandel Coast' — the *Maranui*

'I still remember the cries they sent out when they were sitting on the raft and we were trying to pick them up,' Swedish seafarer Captain Thoby Lindbergh recalled 30 years later. 'We tried many times to get to the raft, but the last time I saw it there were four people on board and they were so numb from the cold they didn't even reach out for ropes.'[28]

The shipwrecked sailors Lindbergh failed to recover were from the Northern Steam Ship Company's motor vessel *Maranui* in 1968. The *Maranui*, the first coaster built for Northern in 25 years, was delivered by the Dutch shipyard J. Bodewes of Hoogezand in 1953. The Dutch shipyard was an interesting place. Unlike the typical British shipyard sandwiched in the middle of grimy industrial cities, the Bodewes yards — there were 32 of them! — stretched along a leafy 15-km rural canal. The shipyard foreman knocked off work every day at 4.30 to milk his cows, which grazed in a paddock beside his house at the edge of the shipyard.[29]

An underwater photograph of the *Kaitawa* showing a close-up of the damage to the bottom of the hull.
Gift of Wellington Museums Trust, New Zealand Maritime Museum (2012.0.1576)

RANGITANE The trio of the *Rangitiki*, *Rangitata* and *Rangitane* was conceived of in 1925 for the New Zealand Shipping Company's New Zealand–United Kingdom service, a route all three plied between 1929 and the outbreak of the Second World War. Unlike her sister ships, however, the *Rangitane* would not survive the war, with her loss to the German raiders *Orion* and *Komet* making her the largest passenger liner sunk by surface raiders during the war.

LENGTH 160 m
GROSS TONNAGE 16,712
BUILT John Brown & Company, Clydebank, Scotland, 1929

PURIRI HMS *Puriri* was a converted coaster, having been commissioned into the 25th Minesweeping Flotilla in April 1941. It was operating with another minesweeper, HMS Gale, off Bream Head in the northern approaches to the Hauraki Gulf when it struck a contact mine – part of the 228-mine barrage laid in June 1940 by the German raider *Orion* that had earlier claimed the liner *Niagara*. The *Puriri* would be the only naval loss to enemy action in New Zealand waters in the Second World War.

LENGTH 57 m
GROSS TONNAGE 927
BUILT Henry Robb Ltd, Leith, Scotland, 1938

HAUTAPU Famed more recently for its undignified end at the hands of 'a couple of water front rogues' trying to salvage brass from the old hulk, the *Hautapu* had in an earlier life been in the thick of the RNZ Navy's only mutiny. The main cause was the poor rates of pay compared to the rest of the New Zealand Defence Force. The first mutiny, at Devonport, took the form of a series of non-violent mutinies among the enlisted sailors of four ships and two shore bases. Sailors on the *Hautapu* became involved after the first mutiny broke out: they went on strike to express their dissatisfaction with the government's response to the strikers' demands. The majority of the *Hautapu* mutineers were charged with mutiny and desertion, though were treated leniently in their sentencing. In reality the mutinies were a major blow to the Navy, with around 20 percent of its sailors either discharged or otherwise punished for their actions.

LENGTH 41 m
GROSS TONNAGE 447
BUILT Stevenson & Cook, Port Chalmers, 1942

KAITAWA Lost with all hands in heavy seas in May 1966, the *Kaitawa* was one of eight similar AC Class vessels — six built in Leith and two in Australia — in the Union Company fleet. At times critiqued for the loss of revenue these small and slow ships racked up owing to their limitations in bad weather, the 'slow greens', as they were colloquially known, were specifically designed for their shallow draft, making them particularly suitable for the West Coast bar harbours that were central to the New Zealand coal trade.

LENGTH 85 m
GROSS TONNAGE 2485
BUILT Henry Robb Ltd, Leith, Scotland, 1949

RAINBOW WARRIOR The state-sponsored act of terrorism that was the *Rainbow Warrior* bombing would be a defining moment in New Zealand foreign policy, and a galvanising moment in popular support for the country's anti-nuclear movement. Her final resting place is the Cavalli Islands, where she was scuttled off Motutapere Island on 12 December 1987.

LENGTH 40 m
GROSS TONNAGE 418
BUILT Hall, Russell & Company, Aberdeen, 1954

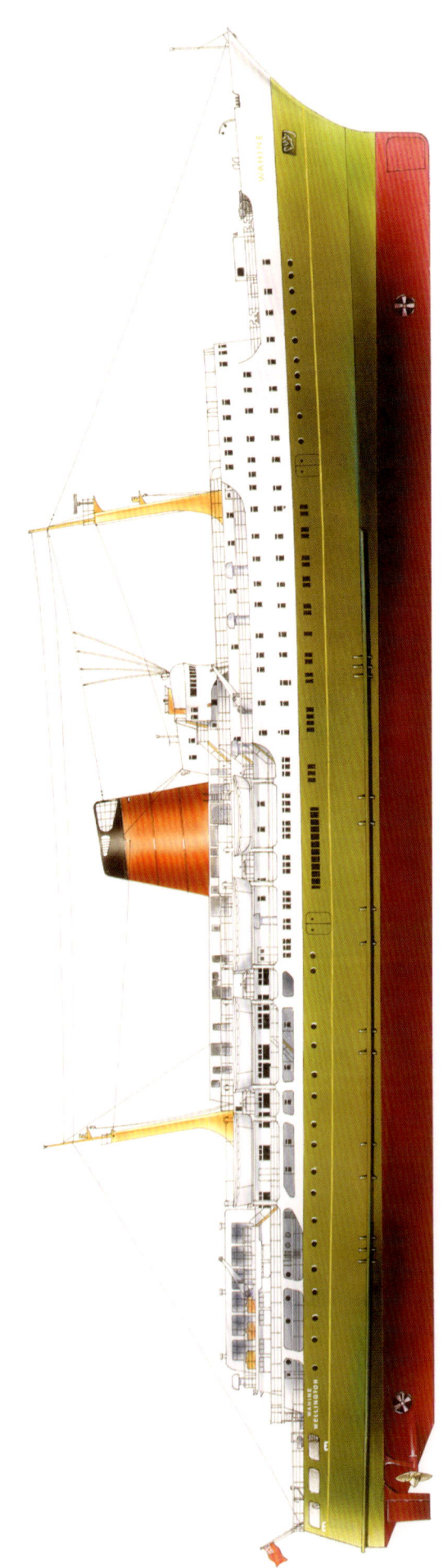

WAHINE The *Wahine* disaster holds an almost mythical place in New Zealand 1960s popular culture. Yet the disaster's significance goes well beyond the events of the day, with the tragedy directly leading to improved safety procedures on ships and the creation of two significant rescue services: the Wellington Volunteer Coastguard, which launched its first vessel within a year; and the Life Flight Trust. The seed for an air rescue service had been planted in Peter Dutton's mind in response to the challenge Wahine rescuers had faced, and with the help of Wellington businessman Mark Dunajtschik, and neurosurgeon Russell Worth, the Life Flight Trust was officially launched in 1976. Beyond this, the root cause of the *Wahine*'s loss — Cyclone Giselle — also triggered the instigation of mandatory civil defence plans by local authorities.

LENGTH 149 m
GROSS TONNAGE 8948
BUILT Fairfield Shipbuilding and Engineering Company, Glasgow, Scotland, 1965

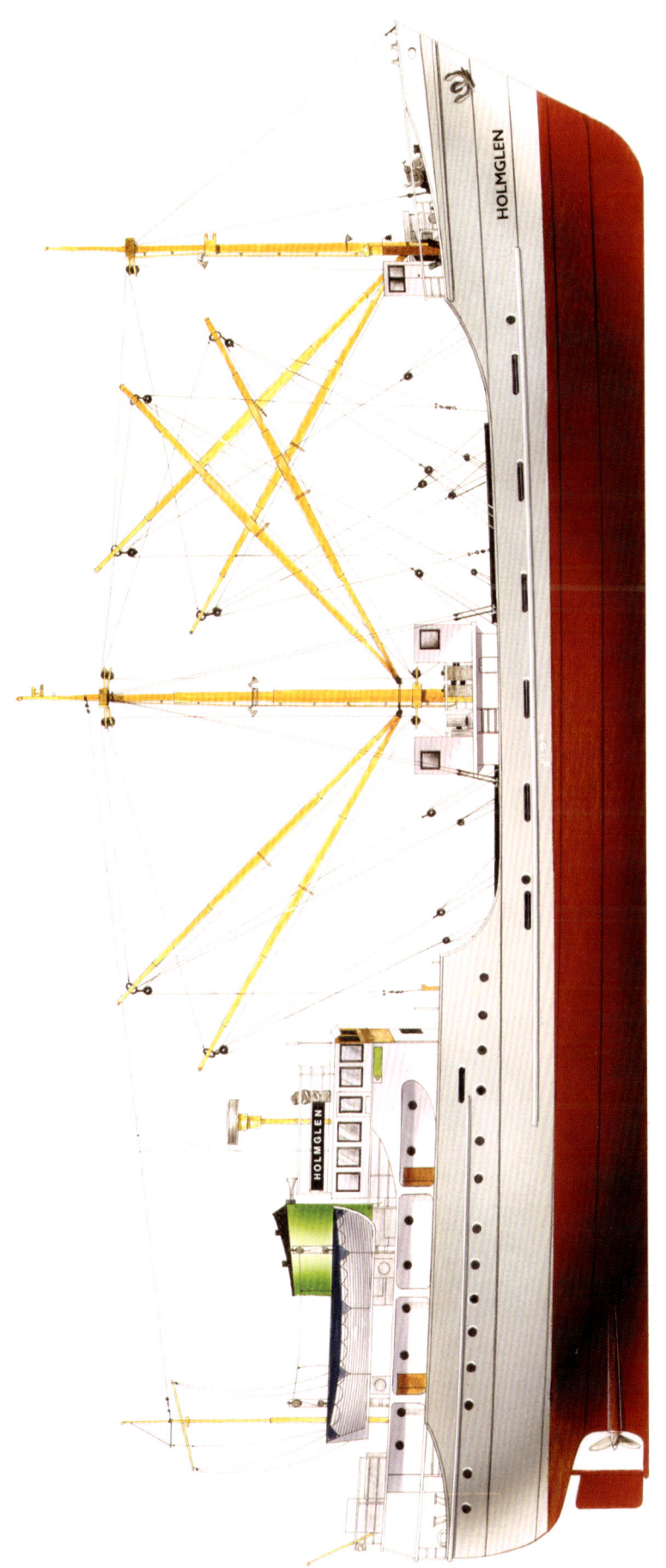

HOLMGLEN Despite MV *Holmglen* being a mere three years old and one of the most modern coasters operating in New Zealand at the time of her loss, the actual cause of her sinking remains a mystery. Speculation was rife, but a court of inquiry could find no explanation for the sinking.

LENGTH 45 m
GROSS TONNAGE 485
BUILT Maartenshoek, Netherlands-based shipyard Bodewes Scheepswerven, 1956

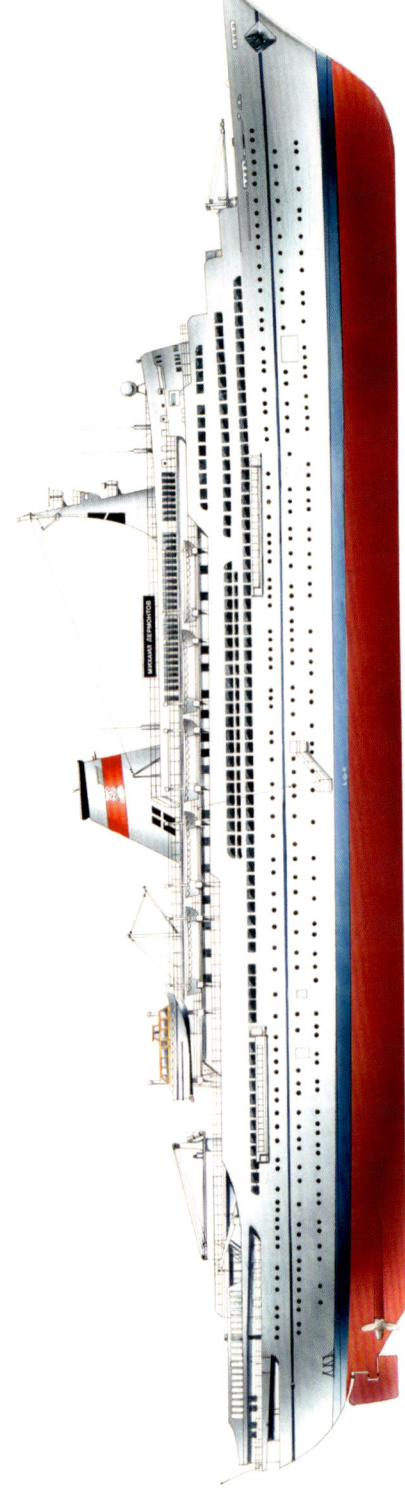

MIKHAIL LERMONTOV The sinking of the *Mikhail Lermontov* in 1986 was the start of an unusually tragic year for the Russian merchant marine. The *Lermontov* was the nation's most luxurious cruise liner and pride of the merchant marine. While only one life was lost, six months later the balance sheet was inverted when the veteran Soviet liner *Admiral Nakhimov* collided with a bulk carrier in the Black Sea, sinking in eight minutes at the cost of 423 lives.

LENGTH 155 m
GROSS TONNAGE 20,027
BUILT VEB Mathias-Thesen Werft, Wismar, in the former German Democratic Republic, 1972

LADY ELIZABETH II Wellington's first purpose-built patrol launch, the *Lady Elizabeth II* entered service in 1973 to replace the original *Lady Elizabeth*, which itself had entered service in 1941 after being requisitioned under the Shipping Requisitioning Emergency Regulations when New Zealand joined the Second World War. Responsible for saving hundreds of people and millions of dollars' worth of shipping in her thirteen-year life, the *Lady Elizabeth II* was lost in July 1986 when the vessel was overpowered by mountainous seas surging through the harbour entrance and capsized. In what is arguably one of the most famous photographs from 1980s New Zealand, two of the four crew were rescued by the actions of Westpac Rescue Helicopter pilots Peter Button and his son Clive — often flying around and below the level of the waves.

LENGTH 15 m
GROSS TONNAGE 24
BUILT Auckland, 1972

Launching the Northern Company's new coaster *Maranui* at Hoogezand in the Netherlands. The *Maranui* was one of several modern motor ships built for the Northern Company to break into longer distance interisland trading.
Photograph by Piet Boonstra. Northern Steam Ship Company archives, New Zealand Maritime Museum (1987.369.7)

The canal was also so narrow that the *Maranui* was launched sideways.[30]

The *Maranui* initially entered Northern's general cargo trade between Whangarei, Tauranga and South Island east coast ports. By 1967, however, the Cook Strait rail ferries were causing the company serious problems and that year the *Maranui* was used to trial small containers and pallets. According to Northern's traffic manager, they enabled the ship to complete six voyages a month instead of four.[31] But the defection of general cargo to the ferries meant that the vessel spent most of its time carrying bulk grain — as did the *Bay Fisher* and *Poranui* — and the *Maranui* loading wheat became a familiar sight at Lyttelton, Oamaru and Timaru.

On 10 June the *Maranui* cleared Lyttelton at 1830 hours, bound for Auckland with approximately 900 tonnes of bulk wheat. All went well until it passed Cook Strait the next day and strong winds and heavy seas started to shift the cargo. With a list of about five degrees, the ship continued up the coast.

By the 13th the *Maranui* was off the Bay of Plenty, and between the Mercury Islands and the Coromandel Peninsula heavy seas increased its list. At 1125 hours it had increased to ten degrees; by early afternoon the wind had reached storm proportions, and by 1535 the ship was obliged to heave to, by now listing by 35 degrees. Water was finding its way below through vent pipes above the scuppers and through a warped and damaged door under the forecastle door on the port side. A broken hinge-pin thwarted every effort to shut it; heavy seas poured over the forecastle head, flooding the port side of the well deck as the list

The New Zealand Herald

Telephones { Classified Advertising - 34-460 / Other Departments - 34-400 — AUCKLAND, FRIDAY, JUNE 14, 1968 — Price 4c Air Freight 5c

Coaster Feared Sunk: Nine Missing In Gale-Hit Raft

Nine men who abandoned their sinking ship in a hurricane yesterday were drifting helplessly in a rubber raft somewhere off the Coromandel coast early this morning.

After hours of pounding by the hurricane, their ship, the 739-ton wheat coaster Maranui, is thought to have sunk last night 31 miles east of Great Mercury Island.

Six of the 15 men on board were plucked from the raft by the crew of the Swedish freighter Mirrabooka.

The Mirrabooka reported at 11.25 p.m. that she had lost contact with the raft after keeping station with it for five hours. The ship had been unable to secure the raft because it was feared movements between the two could tear a hole in the rubber fabric.

The ship radioed that she would stay in the area throughout the rest of the night and would make a search at first light.

The Maranui herself disappeared from a radar screen on board the Mirrabooka at 6.58 p.m.

Only about 30 minutes earlier, the 15 men on board had abandoned ship and launched a tiny rubber raft into furious seas. A full gale was blowing.

Bulk Wheat

Heavily laden with 930 tons of bulk wheat, the Maranui had been bound for Auckland from Lyttelton when she ran into strong winds as she neared her home port.

Running before the gale, the master, Capt. D. A. Bruce, of 12 Jonathan Place, Campbell's Bay, had trouble maintaining course and about 4.40 p.m. he radioed to shore that he had decided to heave to.

In this position facing the oncoming seas, he hoped to ride it out.

The position of the Maranui was given as 31 miles east of Great Mercury Island—36 degrees 48 minutes south, 176 degrees 25 minutes east.

Winds up to 80 miles an hour seized the ship as she was turning into the sea, and the wheat cargo shifted. The Maranui soon took a heavy list.

Mayday

Auckland Marine Radio picked up Mayday signals from the coaster at 5.03. The Maranui reported she was listing badly and in urgent need of help.

A general alert was flashed to shipping in the area.

The Maranui was heeling at about 35 degrees to port. She reported that her engines and steering were useless and that she could not launch her two lifeboats.

The order was given to abandon ship and at 6.23 p.m. the 15-man crew took to the rubber raft. The raft, which can carry 15 men, was equipped with six red flares and smoke flates. It carried a red light.

Stood Off

One of the first to respond to the Mayday call was the Mirrabooka. She signalled that she was in the area on her way from Napier to Auckland to finish loading a general cargo and wool for European ports.

She hurried to aid the Maranui and, guided by flares, stood off the listing coaster at 6.15 p.m. just before the crew abandoned ship.

Shore stations waited tensely for word of the crew as they tossed helplessly in vicious seas. The Mirrabooka pumped out oil to try to stop the waves breaking and make rescue work easier.

What had been a gale lashed itself into a hurricane as the night progressed.

At hurricane force—12 on the Beaufort scale—the air is filled with foam and spray. Visibility is only a few yards. Waves can reach 45 feet.

At 8.34 p.m. the Mirrabooka radioed that the lights of the raft were still in view.

Six survivors were on board the rescue ship by 9.19 p.m.

A Search and Rescue centre was set up at Whenuapai airfield at 7 p.m. Mr T. Keal-Smith, of the Department of Civil Aviation, was Search and Rescue controller.

More than 50 people helped to plot rescue moves at the centre.

An 11-man Orion aircraft crew was put on the alert in case the Mirrabooka could not find the liferaft. The crew was stood down later in the evening until daylight.

It is thought that most of the crew come from Auckland. Many relatives called the offices of the owners, the Northern Steam Ship Co., Ltd., when they heard the Maranui was in danger.

The police said the six men taken on board the rescue ship are Messrs R. J. Ingham, C. J. Taylor, G. B. Monk, E. Hampson, F. McHardy and J. E. Cameron.

The Maranui was built in Holland in 1953 and was converted for carrying South Island wheat cargoes last year.

With another of the Northern Steam Ship Company's ships, the Maunganui, she pioneered the southern coastal wheat trade.

International regulations stipulate the shifting board arrangements that have to be rigged before a freighter can carry bulk wheat. In heavy weather a wheat cargo, if unchecked, has a tendency to shift causing a ship to list.

The shifting boards fitted to most freighters consist of a vertical timber bulkhead running down the centre line of a cargo hold in a fore and aft direction.

The timber bulkhead acts rather like sectional baffles in a petrol tanker.

DISTRESSED YACHT SAFE AFTER LONG ORDEAL

The Christina arriving at Russell yesterday after being towed in by the fisheries protection vessel Mako.

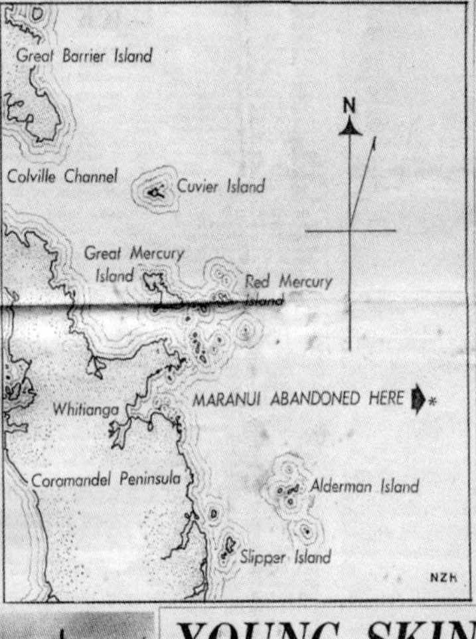

Pirate Radio Ship In Trouble

The pirate radio ship Tiri II struck trouble again last night in high winds and rough seas.

Radio Hauraki went off the air about 9.30 p.m.

A message from Tiri II just after midnight said that if present weather conditions continued it would probably need a tug.

It was hoped the drifting boat would pass to the north-west of Little Barrier Island.

With the Tiri II were Mr Bill Gibb and his launch, Marauder.

The ship had lost steerage and was making no headway, a Radio Hauraki director, Mr Derek Lowe, who was in contact with the Tiri II through Auckland Radio, said at 10.30 p.m.

One of her four engines had also cut out.

The decks were awash but the vessel was reported to be "all tight below."

Mr Lowe said the Tiri's transmitter was not working and telephone contact had been severed when lines on Great Barrier Island came down.

The naval tug Arataki was expected to leave early this morning to go to the aid of the Tiri.

The Maranui during a survey on the Auckland Harbour Board slip.

Lachlan Has To Run For Shelter

After making a resolute effort to reach the Maranui rescue scene through the hurricane, the survey ship HMNZS Lachlan had to run for the shelter of the Coromandel coast shortly after one o'clock this morning.

One blast of wind which rocked her was measured at 104 miles an hour.

The Lachlan was notified of the Maranui's position about 5.15 p.m. She was then about 40 miles away.

Under Cdr I. S. Monro, she was hove to 17 miles from the Mirrabooka about 12.35 a.m. She was making only two knots in waves up to 45 feet.

The hurricane was centred about 100 miles north-east of Great Barrier Island last night and slowly moving south-east.

A weather office spokesman said: "We don't think it will be as bad as the Wahine storm, but it is of the same breed."

The tug Mohala left Auckland at 10.45 p.m. to join the Mirrabooka. It was not known what time she was expected to arrive.

Winds up to 90 knots were recorded at the Mokohinau light, in the northern Hauraki Gulf, last night and many Northland towns were buffeted. Auckland escaped harm.

Floodwaters engulfed several low-lying streets in the Whangarei business area about 10 p.m.

With high tide at 10.29 p.m. water was up to two feet deep in some streets. Many business premises were awash.

New Engine Failed

TONGAN YACHT WRECKED

Four Tongans who conquered a sea of troubles to get a new engine for their boat, were wrecked yesterday when the engine failed to start after a storm blew their sails out.

Twelve Tongans sailed to New Zealand over three months ago, intending to save up for their new engine by working.

They landed at Gisborne, where work was plentiful—but the Ministry of Labour would not allow them to be employed.

Eight returned discouraged to Tonga.

Finally, after many local residents had rallied to their support, the other four found the funds. The engine was installed and they set off for home.

Yesterday their 45-foot boat the Siatukimoana was being pounded on Ocean Beach, north of the Whangarei Harbour entrance, after capsizing.

The men had to swim ashore through thundering surf.

The captain, Mr Malakai Tapealava, said he thought their last moments had come when the boat capsized about a quarter of a mile off shore.

"We hadn't had any sleep or food for two days and the storm was getting worse and worse," he said.

"We sailed past Whangarei and right into the peak of the storm.

"All the rigging blew down and we fixed up a small sail to try to keep the boat steady, but the weather worsened and we decided to beach the boat.

"We were all thrown overboard and started swimming.

"Then the boat rolled right over on top of me. I really thought I was done for. But I came up and saw the other three swimming toward the shore."

It took the four half-an-hour to swim 300 to 400 yards through the surf.

Meal and Shower

Then they staggered about half a mile to Mr I. S. Smith's home on a hill overlooking the beach.

The vessel and the men's belongings were not insured.

Later in the day the Tongans, with the help of five local farmers, used three tractors and a crawler in several attempts to right the vessel.

By late yesterday afternoon they had had no success. Heavy rain and gale-force winds persisted.

Mayday Call

While the Tongan yacht was being washed on to the beach a tense 17-hour mission to rescue another was in progress further up the coast.

About 8.30 on Wednesday night the yacht Christina sent out a Mayday call giving her position as 10 miles north of Cape Brett.

Her skipper, an American, Mr J. Edwards, reported that his 30-foot boat had all its sails blown away except for a storm jib and was being forced inshore by a strong north-easterly wind.

The Navy fisheries protection launch Mako left Russell at 9 p.m. to find the yacht. After some difficulty she sighted the Christina a little after midnight about four miles from the Cavalli Islands.

The wind had gone round to the south-east and was blowing strongly and the Christina was not taken in tow until 4 a.m.

With a gale warning out, the Mako began a long slow tow back to Russell, averaging about three knots.

Towline Broke

Waves were towering above the Mako's 28-foot mast and about 9 a.m. the towline parted as the two vessels neared Tikitiki Island, at the entrance to the Bay of Islands. It took an hour and a half to get under way again.

The Christina was finally anchored off Russell at 2 p.m.

Her owner, who, with his New Zealand wife, had set off for the New Hebrides from Russell on June 5, said he had been caught in the storm and driven back many miles toward New Zealand.

All but one of the sails had been ripped to shreds and the engine would not start because water had been forced into it through the exhaust pipe.

The hull of the stoutly-built Christina suffered only minor damage, although a dinghy and raft were washed from her deck.

YOUNG SKINDIVER SAVES TWO

A young skindiver made two hazardous swims from the shore to Goat Island, near Leigh, yesterday to rescue two fishermen marooned there after their boat had broken up.

The two men, Mr Willem Post, aged 50, and Mr Marinus Kragt, aged 60, both of Leigh and co-owners of the 32-foot fishing launch Sea Wind, were finally brought to safety at 4.40 p.m.

Their rescuer was 21-year-old Mr A. M. Ayling, of Green Bay, Auckland.

After swimming to Goat Island yesterday morning Mr Ayling made another crossing late in the afternoon to take over a line.

By Dinghy

By means of a pulley a dinghy was taken across and the two men were hauled back in it.

Mr Ayling and Mr R. Snowball, a friend, dragged themselves back along the line.

The drama began early yesterday when the two men were returning from fishing. The Sea Wind went aground on Goat Island when the motor failed.

One of the men was seen briefly about 8 a.m. Then he disappeared.

An air force Iroquois helicopter circled the island three times without sighting the men. High winds eventually forced it to return to Whenuapai.

Mr Ayling said he put on his skindiving gear—wetsuit, flippers, snorkel and mask—and swam the 200 yards from the shore to Goat Island about 12 o'clock.

"I went across to see if anyone needed help," he said. "They had built a sort of hut on the island.

"When I got there they were pretty pleased to see me. They were a bit scratched and had on only pyjamas and parkas.

"They said the motor of the Sea Wind had failed. They threw out a couple of anchors but they did not hold. They went aground on the point of the island where the waves are biggest."

Mr Ayling said that between 8 a.m.—when the Sea Wind went aground—and late in the morning the launch had pounded on the rocks and smashed to bits.

The two men had told him they felt lucky to be alive.

It would have been impossible for the men to have swum to shore earlier in the day because of extremely heavy seas which were running up to 10 feet.

Mr Ayling said that even in skindiving gear the swim across had been very difficult.

"If anything," he said, "the seas were even heavier the second time I swam across."

Mr Ayling is a student at the University of Auckland Marine Research Laboratory at Leigh. He has been skindiving for about seven years.

Mr Post's brother, Mr J. Post, said his brother had bought the Sea Wind only in April. It was valued at more than $10,000.

The two fishermen had moved to Leigh about six weeks ago.

Veteran VC's Flight Hitch

MAY MISS REUNION

Unless last-minute arrangements are made, Mr J. G. Grant, an Auckland Victoria Cross veteran of the First World War, will not be able to attend the reunion of VC winners in Britain next month.

Mr Grant, aged 79, a resident of Roskill Masonic Village, is a semi-invalid.

He had hoped to attend the reunion to be given by the Queen at Windsor Castle, with the help of the matron of the village, Sister Marjorie Dalton, and Mrs Barbara Fair, a nurse.

Each VC is allowed to take two relatives and Mr Grant was given permission to take the two village staff members.

The trio had their personal invitations from the Queen and had received inoculations and passports in preparation for the trip.

Then came a hitch. The Royal New Zealand Air Force, which was going to fly the New Zealand contingent to Singapore to connect with a Royal Air Force plane, said it could take Mr Grant but not his nurse aides.

There was even some doubt that the RNZAF could take any of the New Zealand VCs.

Group Capt. R. K. Orrock, of the British High Commissioner's staff in Wellington, has now arranged for an RAF Air Support Command plane to carry the VCs and their relatives from New Zealand to Britain.

Unfortunately, because of the route the plane will be flying—by way of India and the Mediterranean—the RAF has declined to take Mr Grant.

It is thought that the trip would be too arduous for him.

The plane, probably a Comet jet aircraft, is expected in New Zealand about July 9. About four or five VCs, each with two relatives, will make the trip to Britain.

Group Capt. Orrock said it was hoped that Capt. C. H. Upham, the double VC winner, would be in the party, but it was not yet known whether he would be able to go.

Offers have been made to fly Mr Grant and Sister Dalton and Nurse Fair to Britain by a commercial airline so the veteran VC may yet meet the Queen.

TWO BREAK GAOL

Press Assn. — New Plymouth

Two men—one with a conviction for aggravated assault—escaped over the wall of New Plymouth Prison yesterday afternoon. They were still at large last night.

ROBBED BY BOGUS WORKMEN

Two young men, posing as council workmen, robbed a woman pensioner of about $10 at her flat in Grey Lynn yesterday morning.

Miss Florence Head, aged 84, who lives in a block of Auckland City Council pensioner flats in Surrey Crescent, said the men, both dressed in overalls, told her they had come to check the water service.

Miss Head said she did not think anything more about it until she went to get her purse. It was gone.

Buckland's Ticket Wins $60,000

NZPA-Reuter — Melbourne

A New Zealand syndicate won first prize of $60,000 in a lottery drawn in Melbourne yesterday.

The prize went to ticket No. 39201 in the name of Kelly's Eye Syndicate, Buckland's Beach, Auckland.

GIVEN SHEEP'S HEART

NZPA-Reuter — Houston, Texas

Surgeons at the St Luke's Episcopal Hospital transplanted a sheep's heart into a man, the hospital announced last night.

The hospital administrator, Mr Newell France, said the operation was intended to keep a 47-year-old seriously ill man alive until a human heart donor was available, but he died soon after the operation.

It was the second operation involving the transplanting of an animal heart into a human.

According to Mr France, the team, which had carried out four heart transplants in recent weeks, decided to perform the operation after the man's heart stopped and resuscitation attempts failed.

The heart was taken from a 120-pound ram but soon after it was sewn into place the man died.

Of the four previous transplants at St Luke's, two recipients died. Two are recovering and are in a satisfactory condition. They are 47-year-old Mr Everett Thomas, of Phoenix, Arizona, who received a new heart on May 3, and 54-year-old Mr Louis John Fierro, of Elmont, New York, who underwent a transplant operation on May 21.

In Cape Town, Dr Philip Blaiberg was maintaining his improvement after a sudden relapse, Professor Christiaan Barnard said last night.

Dr Blaiberg has been fighting hepatitis since Monday.

Professor Barnard examined Dr Blaiberg for more than half an hour in the intensive care unit where the 59-year-old dentist has been since his relapse.

Professor Barnard refused to answer reporters' questions about Dr Blaiberg's chances of survival.

A message from London says that Mr Frederick West, the first heart transplant patient in Britain, still has kidney trouble, but his general condition was better last night.

The latest bulletin on 45-year-old Mr West said: "Although his kidneys have not yet responded to supportive measures, Mr West's general condition shows improvement. His heart is working well."

increased. The *Maranui*'s stability was being affected as water sloshed about its double bottom; the cargo, unrestrained by shifting boards, was moving, adding to the crew's fears.

At 1703 hours the ship issued the first Mayday call. Seafarer Eric Hampson later recalled being on the bridge and hearing the mate ask, 'is this a Mayday?' and getting the response from Captain Bruce: 'What the bloody hell do you think it is?'[32] Two ships responded, HMNZS *Lachlan,* sheltering off Great Mercury Island, and the Swedish freighter *Mirrabooka*. Although the naval vessel did its best, strong head winds and boisterous seas (up to 12 metres high) forced it to heave to with weather damage that later obliged it to limp away for repairs. That left only the *Mirrabooka,* then a few miles away. At 1742 it established visual contact with the coaster.

Thirty-three minutes later Captain David Bruce ordered his crew to abandon ship. Although Bruce had swung out the ship's two lifeboats, the weather rendered them useless; only the inflatable life raft could be launched, and it was a precarious perch. 'We'd never seen a life raft open until we had to jump into it during the storm,' Graeme Monk recalled. 'The first things we found when we opened it up were two paddles — about as useful as ping-pong bats when you're in 21-metre waves — and a kitchen sponge to mop up the water in the raft, which was a joke because it was up to our chests.'[33] For the next two hours it drifted around out of sight of the nearby *Mirrabooka,* which did not sight the liferaft until 2019 hours.

Captain Thorsten Wahlstedt manoeuvred his big ship within reach. It was not easy, and his first four attempts failed. He succeeded on the fifth but only six men — chief officer R.C. Ingham, second officer C.J. Taylor, ABs F. Hampson and G.B. Monk, OS F. McHardy and cook J.E. Cameron — managed to make the hazardous ascent up the ropes and nets strung along the Swede's side. One man was dashed against the *Mirrabooka*'s high side (unfortunately, the ship was lightly laden and therefore high in the water) and then swept away; two or three others were seen to be carried away before the raft, still containing four men, disappeared from view. It was next seen on 16 June at the base of a cliff on Great Mercury Island.

No more survivors were found and only one body was recovered, by the crew of HMNZS *Kahawai*. Those lost were the master, David Bruce, chief engineer M.C. O'Flaherty, second engineer R.G. Watson, third engineer J. Walton, boatswains S.C. Henry, R.E. Orr and L.S. St Bruno, wiper J. H. McPherson and steward J.C. Roberts.

The actions of the *Mirrabooka*'s crew were widely praised. 'In the real-life drama enacted off the Coromandel coast the *Mirrabooka* answered the call not just in a storm but in a virtual hurricane — and in inky darkness into the bargain,' the *New Zealand Herald* said. 'In such conditions, the preservation of the safety of their own ship must alone have been an ordeal for Captain Wahlstedt and his men.'[35] The Minister of Marine sent Cabinet's thanks to Captain Wahlstedt.[36]

The court of inquiry found that the ship had sunk after the cargo of wheat shifted. A contributing cause was the entry of water into some of the *Maranui*'s double bottom tanks on the port side through uncovered and unplugged air pipes. The court criticised the master and the chief officer for omitting to plug the air pipes and for running a loose ship; it was also unimpressed by the Northern Company's joint general manager and its marine superintendent. As a result the *Poranui* and *Bay Fisher* were fitted with shifting boards.

OPPOSITE All at sea. The *New Zealand Herald*'s front page attested to a wild and windy day in the South Pacific.
New Zealand Herald

14

Close calls — successful salvages

Wellingtonians often use ship names to describe their weather. Whenever a gale lifts roofs and waves crash across roads, they talk about '*Wahine* weather'. Many older residents will call a long, calm, sunny patch of weather as '*Wanganella* weather'. Both phenomena take their names from ships snagged by Barrett Reef.

The *Wanganella* was a popular trans-Tasman liner owned by Huddart Parker Ltd of Melbourne. It had been ordered as the *Achimota* for Elder Dempster but was sold before completion to Huddart Parker in 1932. With its boxy lines and squat funnels, the 9576-ton *Wanganella* carried 304 first-class and 104 second-class passengers.[1] It served as a hospital ship during the war and spent most of 1946 in a shipyard being refurbished for a return to civilian service. Painters were still aboard the fully booked ship when it left Sydney late in December that year.

Just before midnight on 19 January 1947, the *Wanganella* struck the edge of Barrett Reef. The reef — an old enemy to shipping — is almost a kilometre long. Most of the rocks are visible and run alongside the main shipping channel. After the *Devon* (see pages 150–54) was lost in 1913 by being over-cautious in avoiding Barrett Reef, a light buoy had been placed on the reef's seaward end to warn mariners.[2] On what was now a very dark 1947 night, Captain R. Darroch appeared to have mistaken the seaward light for a pile beacon. Fortunately, the sea was calm and the ship held onto the reef instead of sliding off and sinking. The ferry *Cobar* came alongside the next morning and took off the passengers and most of the crew. Although many people had been shaken out of their bunks by the impact, there were no serious injuries. In fact, some mistook the bump for the vessel berthing at the wharf.

Luck stayed with the *Wanganella*. One strong southerly could have wrecked the ship, but Wellington's weather remained unusually fine for the 18 days that it lay on the reef. Salvors sealed off the ship's damaged forepart and pumped it out. On the night of 6/7 February a strong southerly swell and some solid pulling from the tugs lifted it off the rocks. A salvor rushed from the dining room to break a bottle of Australian beer against its hull to celebrate. The tugs towed

OPPOSITE The Liberty ship *Viggo Hansteen* aground off Tikoraki Point near Moeraki on 24 April 1952. Thick seaweed cushioned the impact and a calm sea enabled the ship to be refloated undamaged.
Otago Harbour Board.

An aerial view of Huddart Parker's trans-Tasman liner *Wanganella* aground on Barrett Reef near the entrance to Wellington Harbour.
Ref: EP-Ships-Wanganella-01, Alexander Turnbull Museum, Wellington, NZ

the *Wanganella* stern-first to a wharf where it settled on the bottom. The ship entered the Jubilee floating dock on 18 February after shipwrights had reduced its draught forward.

There the ship's luck ran out again when the shipwrights walked off the job before much work had been done. The *Wanganella* was taken out of the floating dock and laid up at Clyde Quay Wharf bow-down in the mud – ultimately not returning to service until November 1948, almost two years after striking Barrett Reef. The court of inquiry, held over three days in February 1947, blamed Captain Darroch for mistaking the Barrett Reef floating buoy for the No. 1 leading light and suspended his certificate for three months.

Within another decade airline competition was strangling trans-Tasman shipping. In September 1961 Huddart Parker sold its shipping interests to Mcilwraith McEacharn, which kept the old ship going until April 1962 when it was sold to a Hong Kong company. A year later it was back in New Zealand at Deep Cove, Fiordland, as a hostel ship for workers on the Manapouri power project. Its engines silent, the *Wanganella* housed construction workers until 1970 when it was scrapped.

Taking a liberty at Moeraki

The residents of Moeraki, North Otago — a port that had not seen a ship for nearly 70 years — awoke with a shock on the morning of 24 April 1952 to see an ocean-going freighter ablaze with lights just metres from the shore off Tikoraki Point.[3] It was the Norwegian Liberty ship *Viggo Hansteen*. Launched in Baltimore in 1943 as the *George M. Shriver* but renamed in honour of a Norwegian patriot executed by the Germans in 1940, it was travelling from London to Dunedin. Luck had deserted it a long time ago. It had had only two days of fine weather during its voyage halfway around the world.

Lacking radar and unfamiliar with local waters, the *Viggo Hansteen*'s crew mistook Moeraki's lighthouse for Taiaroa Head at the entrance to Otago Harbour. The lookouts stationed in the bow took alarm only when the ship ground across the rocks. Fortunately the thick kelp beds cushioned the impact and kept damage to a minimum. One hundred and forty metres offshore and in just five metres of water, the *Viggo Hansteen*'s position was more humiliating than life-threatening, especially since the sea was calm.

The Otago Harbour Board sent the tug *Dunedin,* which by 1520 hours had the ship off the rocks and was escorting it to Otago Harbour. A docking at Port Chalmers revealed no damage and

14

Close calls — successful salvages

Wellingtonians often use ship names to describe their weather. Whenever a gale lifts roofs and waves crash across roads, they talk about *'Wahine* weather'. Many older residents will call a long, calm, sunny patch of weather as *'Wanganella* weather'. Both phenomena take their names from ships snagged by Barrett Reef.

The *Wanganella* was a popular trans-Tasman liner owned by Huddart Parker Ltd of Melbourne. It had been ordered as the *Achimota* for Elder Dempster but was sold before completion to Huddart Parker in 1932. With its boxy lines and squat funnels, the 9576-ton *Wanganella* carried 304 first-class and 104 second-class passengers.[1] It served as a hospital ship during the war and spent most of 1946 in a shipyard being refurbished for a return to civilian service. Painters were still aboard the fully booked ship when it left Sydney late in December that year.

Just before midnight on 19 January 1947, the *Wanganella* struck the edge of Barrett Reef. The reef — an old enemy to shipping — is almost a kilometre long. Most of the rocks are visible and run alongside the main shipping channel. After the *Devon* (see pages 150–54) was lost in 1913 by being over-cautious in avoiding Barrett Reef, a light buoy had been placed on the reef's seaward end to warn mariners.[2] On what was now a very dark 1947 night, Captain R. Darroch appeared to have mistaken the seaward light for a pile beacon. Fortunately, the sea was calm and the ship held onto the reef instead of sliding off and sinking. The ferry *Cobar* came alongside the next morning and took off the passengers and most of the crew. Although many people had been shaken out of their bunks by the impact, there were no serious injuries. In fact, some mistook the bump for the vessel berthing at the wharf.

Luck stayed with the *Wanganella*. One strong southerly could have wrecked the ship, but Wellington's weather remained unusually fine for the 18 days that it lay on the reef. Salvors sealed off the ship's damaged forepart and pumped it out. On the night of 6/7 February a strong southerly swell and some solid pulling from the tugs lifted it off the rocks. A salvor rushed from the dining room to break a bottle of Australian beer against its hull to celebrate. The tugs towed

OPPOSITE The Liberty ship *Viggo Hansteen* aground off Tikoraki Point near Moeraki on 24 April 1952. Thick seaweed cushioned the impact and a calm sea enabled the ship to be refloated undamaged.
Otago Harbour Board.

An aerial view of Huddart Parker's trans-Tasman liner *Wanganella* aground on Barrett Reef near the entrance to Wellington Harbour.
Ref: EP-Ships-Wanganella-01, Alexander Turnbull Museum, Wellington, NZ

the *Wanganella* stern-first to a wharf where it settled on the bottom. The ship entered the Jubilee floating dock on 18 February after shipwrights had reduced its draught forward.

There the ship's luck ran out again when the shipwrights walked off the job before much work had been done. The *Wanganella* was taken out of the floating dock and laid up at Clyde Quay Wharf bow-down in the mud – ultimately not returning to service until November 1948, almost two years after striking Barrett Reef. The court of inquiry, held over three days in February 1947, blamed Captain Darroch for mistaking the Barrett Reef floating buoy for the No. 1 leading light and suspended his certificate for three months.

Within another decade airline competition was strangling trans-Tasman shipping. In September 1961 Huddart Parker sold its shipping interests to Mcilwraith McEacharn, which kept the old ship going until April 1962 when it was sold to a Hong Kong company. A year later it was back in New Zealand at Deep Cove, Fiordland, as a hostel ship for workers on the Manapouri power project. Its engines silent, the *Wanganella* housed construction workers until 1970 when it was scrapped.

Taking a liberty at Moeraki

The residents of Moeraki, North Otago — a port that had not seen a ship for nearly 70 years — awoke with a shock on the morning of 24 April 1952 to see an ocean-going freighter ablaze with lights just metres from the shore off Tikoraki Point.[3] It was the Norwegian Liberty ship *Viggo Hansteen*. Launched in Baltimore in 1943 as the *George M. Shriver* but renamed in honour of a Norwegian patriot executed by the Germans in 1940, it was travelling from London to Dunedin. Luck had deserted it a long time ago. It had had only two days of fine weather during its voyage halfway around the world.

Lacking radar and unfamiliar with local waters, the *Viggo Hansteen*'s crew mistook Moeraki's lighthouse for Taiaroa Head at the entrance to Otago Harbour. The lookouts stationed in the bow took alarm only when the ship ground across the rocks. Fortunately the thick kelp beds cushioned the impact and kept damage to a minimum. One hundred and forty metres offshore and in just five metres of water, the *Viggo Hansteen*'s position was more humiliating than life-threatening, especially since the sea was calm.

The Otago Harbour Board sent the tug *Dunedin,* which by 1520 hours had the ship off the rocks and was escorting it to Otago Harbour. A docking at Port Chalmers revealed no damage and the ship survived another 12 years until, as the *Alkimos*, it went ashore for good off the Australian coast.

A ship of fools — the *Pacific Charger*

That all the radar, satellite navigation equipment and technology in the world would still not cancel out human stupidity became apparent at 0304 hours on 21 May 1981 when the 10,000-ton bulk carrier *Pacific Charger* hit rocks at Baring Head near the entrance to Wellington Harbour. Although a 'flag of convenience' ship (i.e. one flagged out by its owners onto a register which required cheaper crewing standards and numbers), the *Pacific Charger* was brand-new. The 146-metre ship was no beauty, but it was of a well-proven Japanese design for the flexible handling of log and container cargos.

Launched at the Sasebo Heavy Industries yard on 21 February 1981, the ship had been delivered to its owners, Ocean Chargers Ltd of Monrovia, Liberia — a wholly owned subsidiary of the Kansai Steamship Co. Ltd of Japan — on 24 April. Chartered back by Ocean Chargers' parent company, the bulker had been dispatched to New Zealand on its maiden voyage as part of an agreement with the British Crusader Swire Container Service. Confused? Others certainly were. These multi-level ownership arrangements, common in modern shipping, would give the court of inquiry plenty of detective work. At the time of the stranding the *Pacific Charger* was carrying 4000 tonnes of car components and was under the command of Taiwanese born Captain R.Y. Chiou.

The weather was as bad as Wellington's critics usually imagine it to be; a 70-knot south-easterly gale was shrieking through the strait and visibility was down to less than two kilometres. Alerted by radio, the pilot cutter *Tiakina* and the tug *Kupe* arrived on the scene shortly afterwards but could do nothing since the freighter was a mere 55 metres from the shore and wedged fast. They could not get close enough to do anything. Shore parties fared no better; washouts and slips on the road from Wainuiomata reduced their movement to crawling speed. Not until 1300 hours could the crew abandon ship by dinghy and rope ladder. Cold, wet and miserable, 23 of the 26 Burmese and Taiwanese seafarers were taken to a city hotel by army trucks, leaving the master, chief officer and chief engineer aboard to protect their owners' interests.

Despite being pounded by heavy seas, to which it lay parallel, the *Pacific Charger* showed no signs of breaking up. Although no one had been able to survey the damage accurately, it was obvious that refloating the ship — if indeed that was possible — would be expensive. Kansai gave the salvage job to Singapore-based Captain A.H.G. (Hugh) Murray, whose company, Selco Marine Salvage, signed a standard 'no cure, no pay' agreement and despatched its nearest salvage tug, the *Asiatic Triumph* from Papeete. The *Asiatic Triumph* reached Wellington on the 29th.

By then Murray's teams had accomplished a great deal. The Ministry of Works and Development bulldozed a road from Eastbourne and built a causeway out to the ship, enabling road tankers to remove the *Pacific Charger*'s 800 tonnes of heavy fuel oil, which posed a potential threat to the environment.

By 2 June 1981, the weather had calmed and an attempt was made to refloat the ship. The harbour board tugs *Kupe* and *Toia*, the launch *Tirohia* and the police launch *Lady Elizabeth II* added their mechanical muscle to that of the *Asiatic Triumph*, but to no avail; by 1630, when the attempt was abandoned, the *Pacific Charger* was still stuck. Three days later, Murray tried again. The two local tugs joined the *Asiatic Triumph* to take advantage of high water and a slight southerly swell which, Murray hoped, would help the *Pacific Charger* lift off. This time he succeeded, getting the ship under way by 1935 hours, after which it was towed into a Wellington berth and tied up at King's Wharf in the early hours of the 6th. There it lay, surrounded by a precautionary oil barrier. Selco divers patched up its crumpled hull and had it ready for handing over to its owners on the 17th.

Seven days later shore cranes started removing the oil-covered and water-damaged cargo. On 30 June a steel girder being lifted out of its No. 2 hold slipped from the sling and plunged through the *Pacific Charger*'s hull. Two thousand tonnes of water flooded in until the fire brigade's pumps took over.

The tug *Sumi Maru* arrived on 13 July to tow the *Pacific Charger* to Japan for permanent repairs. A writ of arrest caused a delay, but by the end of the month the legal impediments had been removed and the ship sailed for Sasebo shipyard. It re-emerged later as the *Cobalt Islands*.

The court of inquiry took longer to complete its work than did the shipyard. This one was longer than most, often carried out with the help of translators and was not ready until January 1982. It was almost worth the wait. Court reports do not usually make exciting reading but this one, for all its sombre, legalistic wording, kept the headline writers busy. Its bald answer to the Minister of Transport's question 'what was the cause of the stranding?' provided good copy. 'The vessel was set in by the ebb tide in a south easterly direction, and was being navigated on an unsafe course,' it stated, blaming also 'the inability of the master or second mate to plot the vessel's progress over the last 20 minutes', and adding for good measure, 'the incompetence of the master by his failure to study the tidal information or avail himself of the manuals dealing with that particular area or to even recognise the danger of the situation'.

In blaming this ship of fools, the court spoke bluntly. 'Despite all the international conventions as to safety at sea, it is useless to send a well-found ship on a voyage if the officers and crew are incompetent … Technology did not make human skills redundant.' This was a new ship and none of the officers and crew had sailed together before. There were obvious language problems with a Japanese-owned, Taiwanese-officered and Burmese crewed ship operating along the coast of an English-speaking nation.

Few of the officers emerged with any credit. Chiou, on his first visit to New Zealand and relying on small-scale American charts, was an experienced mariner but had no 'formal qualifications in the use of radar, although he relied on it entirely and alone'. Soe Tint, the quartermaster on duty 'could not be regarded as an experienced quartermaster and his failure to inform the master sooner than he did that he was having difficulty in keeping the ship on a straight course clearly indicates this'. Second mate Kao Hing Ho had never attended navigation school or done radar training: 'he was the navigation officer, but any skill or experience he had in that field was derived only from sea service, most of it in lower ranks'. Third mate Chung Chin Ming had not served at sea for three years before joining the *Pacific Charger*. Only first mate Chang Yung Chin, off duty at the time of the accident, was considered fully competent as a navigator.

The court accepted that weather conditions were bad — the second worst storm since the *Wahine* storm of 1968 — but felt that Chiou could have avoided danger by standing out to sea and waiting for things to improve. But, determined to get into sheltered water to pick up the pilot, and believing there to be shelter just past Baring Head, he pressed on, compounding his problems by reducing speed, since this made a ship of that size harder to handle. Ill-equipped, using inadequate manuals, off course, unaware of the effect of the set and drift of the tide (at least 1.5 knots) on his ship, ignoring the echo-sounder (which would have told him that the water was shallowing) and failing to take a visual compass bearing on the Baring Head light when it became

OPPOSITE A wave breaks over the bow of the *Pacific Charger*. It was no wonder most people initially thought the ship was doomed. *Evening Post*

visible, Chiou was steering straight for land.

There was no excuse for this. Chiou had modern and reliable radar sets and the coastline around Baring Head is a good radar target. His lack of radar training led him astray. Modern radar sets enabled skilled operators to adjust their performance. The GAIN control, as it is called, varies the amplification of the receiver and thus the strength of the echoes as they appeared on the screen. The STC control, better known as the anti-clutter sea control, varies the GAIN control over short distances; it can be used to remove objects behind them. The FTC control is the anticlutter rain, hail or snow control. It can compensate for overuse of STC. Used intelligently, these controls can eliminate confusing weather-related echoes. Twisted too far, on the other hand, they (the STC particularly) can remove everything — including land!

This is precisely what happened. Neither Chiou nor the mate had read the manuals for the set and overdid their use of GAIN and STC without even switching on the FTC, which might have compensated for this. 'There had been no attempt on the voyage from Yokohama to Auckland to correct the charts from Notices to Mariners, and neither the master nor the second mate had bothered to read about the New Zealand coast or harbours in the manuals aboard, apart from the brief reference in *Guide to Port Entry,* and no attempt had been made to consult New Zealand charts,' the court concluded. 'If he had studied the appropriate NZ charts instead of relying on the small scale American charts he was using, he would have realised that the profile of Baring Head is virtually duplicated a short distance inland and he would have been alert to ensure that his observations of target related to the shore line,' it continued. 'His excessive use of the anti-clutter control obliterated the land inside his two-mile variable ring and he had no idea where he was. However, he was so anxious to make his way into the harbour that he ignored the basic rule of seamanship to steer away from possible danger.

He reduced speed, while looking for the pilot, to such an extent that in the weather conditions prevailing, he had no steerage control, the ship just would not respond to the rudder.'

No charges were laid against anyone because the court had no jurisdiction over non-Commonwealth ships. The best it could do was to warn about the danger posed by poorly trained flag of convenience ships. That its concerns were well placed became apparent a few months later. Chiou, given command of another bulk carrier, the *Orient Treasury,* was lost at sea. The *Pacific Charger,* after several changes of name, is still believed to be trading as the *Mandarin Sea.*

Things that go ping in the dark — the *Jody F. Millennium* and the *Tai Ping*

In 2002 two bulk carriers found themselves in big trouble off the coast. The first was the 15,701-ton *Jody F. Millennium,* its ludicrous name a salute to American actor *Jody Foster* and to the year of its construction (2000). Owned by the Twin Bright Shipping Company of Panama and managed by Soki Kisen Co. of Japan, the *Jody F.* was, like the *Pacific Charger,* a modern, well-equipped geared bulk carrier.

The first week of February 2002 found the *Jody F.* at Gisborne, topping off with logs before returning to Japan. On Waitangi Day yet another 'worst storm since the *Wahine* storm' ravaged the lower North Island. At Gisborne — not the world's most sheltered anchorage — heavy swells rolled in, requiring the tugs *Turihaua* and *Titirangi* to hold the bulker against the wharf because its lines were parting and the ship was moving too violently for loading to continue. By evening it had broken several lines. At that point the pilot decided that in view of the danger presented to shore staff, the wharf and the ship, the vessel should leave port.

At 2150, with the ship still inside the breakwater, the pilot disembarked.[4] Two minutes later, however, as it passed the breakwater, the *Jody F.* was struck by the swell and touched bottom in the approach channel, 'slowing and effectively disabling the ship, which was then driven by the sea and swell further on to the swell area north of the channel where it remained hard aground' off Gisborne's Waikanae Beach.[5] The heavy swells (estimated at up to six-metres) had prevented the tugs and the pilot boat from offering help.

The first gobs of thick, tar-like oil were seen heading toward's Gisborne's prized beach on the 9th, by which time a major salvage operation had swung into action. As television crews filmed the big ship wallowing in the waves little more than a stone's throw from the city's prime real estate, United Salvage Pty Ltd of Sydney joined the Maritime Safety Authority and the navy. Their first priority was to remove the *Jody F.*'s heavy bunker oil, which was pumped into inflatable barges and transferred to HMNZS *Endeavour*.

Meanwhile, United Salvage had assembled a small fleet to help lighten the ship. The big offshore supply ship *Pacific Chieftain* was brought in from Taranaki and was joined by the tug *Sea-Tow 22* and barge *Sea-Tow 17*. At first little went right. On 13 February the *Pacific Chieftain* and *Sea-Tow 22* turned the *Jody F.*'s bow about 30 degrees to seaward, but bad weather again intervened. Next day the line from the *Pacific Chieftain* snapped, and the bulker, settled back on the beach near its original position. As the *Pacific Chieftain* was needed at the Maui oilfield, the large tugs *Keera* and *Sea-Tow 25,* were

RIGHT, TOP Gisborne is a tight port to enter and leave. This image shows the *Jody F. Millennium* ashore off the city beach. A helicopter is approaching from astern to lift more logs while two tugs and an anchor handling vessel standoff.
Maurice Affleck, courtesy of Michael Pryce

RIGHT, CENTRE & BELOW Lightening a ship is an important part of refloating any vessel. These photographs show what a massive and expensive job it is, involving chartered barges and helicopters.
Maurice Affleck, courtesy of Michael Pryce

Two jobs were essential if the *Tai Ping* was to be refloated: pumping out ballast and discharging cargo to lighten the ship. Fortunately the vessel was fitted with cranes, making the job easier. Here the local tug *Hauroko* has been joined by the Lyttelton tug *Purau* and Timaru's *Te Maru*.
Chris Howell, courtesy of Michael Pryce

chartered. Meanwhile, 4300 tonnes of logs had been discharged by ship's cranes into the *Sea-Tow 17* and by helicopter, lightening the *Jody F.* sufficiently for the tugs to pull it off on the afternoon of 24 February. After initial inspection at Tauranga, the ship was towed by the *Keera* on the first stage of its journey to Japan for permanent repairs. On 4 September its owners renamed it *Millennium Bright*.[6]

As usual, the MSA and TAIC issued reports. They found the Gisborne District Council (the port owner), Port Gisborne Ltd (the port operator), Adsteam Port Services Ltd (the provider of marine services), the pilot and the ship's master, Captain Joe Dong Seok, all culpable. Acknowledging the severity of the weather, which caused violent surge conditions far in excess of what forecasts suggested could have been produced, the pilot's concerns about harm to shore personnel and damage to the wharf or ship, and the problems associated with alternatives such as keeping the ship berthed; moving to another, possibly more sheltered berth; or trying to hold the ship with tugs, the reports pointed out the basic error — that there was insufficient water in the channel. 'When taking the swell and its effect on the ship into account, the under keel clearance of the ship was probably negative. Simply put, the ship did not have sufficient water to transit the channel at that time.'[7]

Neither the pilot nor the ship's master had adequate information. 'Port Gisborne Ltd failed to dredge and undertake adequate hydrographic surveys to ensure the advertised depths were maintained,' the MSA report noted. 'As a result of these failures Port Gisborne Ltd continued to disseminate unsubstantiated and potentially inaccurate controlling depths for the port.'[8]

Communication between the pilot and the captain was less than perfect. No passage plan for departure was given to the master, who was on his first visit to Gisborne. While cultural differences may have played their part, TAIC criticised the master accepting the advice to depart as an order rather than contestable advice.

The most contentious point was the pilot's decision to leave the ship while well inside the compulsory pilotage district, in fact inside the breakwater. The MSA called this 'dereliction of his duty as a Pilot', dismissing his fear about being over-carried if he was unable to leave the ship once it left port limits.[9] 'The necessity for the pilot to remain on board the vessel was obvious: the departure was in the hours of darkness, there was an extreme swell running accompanied by near gale force winds, the Master had not previously departed the port and it was critical that the vessel remained in the channel.[10] In TAIC's view, had he stayed on board, speed would not have needed to be reduced, giving the ship more forward momentum when it emerged from the shelter of the breakwater. 'It would most probably have still touched bottom but might have had sufficient momentum to continue down the channel. With greater speed, the steering and directional stability of the ship would have been maintained. In addition, had the pilot continued with the control of the ship, continuity would have remained.'[11]

By 2003 Soki Kissen was suing the port company for US$12.5 million and the MSA audited the port company, renamed Eastland Port Ltd, imposing new limits on the navigation of large ships at the port.

A few months later another flag of convenience bulk carrier got into trouble at the other end of the country. While sailing from Bluff for Lyttelton in thick fog on the morning of 8 October, the 16,041-ton Hong Kong-flagged bulk carrier *Tai Ping*, carrying 9500 tonnes of urea fertiliser, ran aground opposite Morrison's Beach on the Tiwai Point side of the port entrance. The ship was stuck fast, and a survey revealed damage to the fore

The first gobs of thick, tar-like oil were seen heading toward's Gisborne's prized beach on the 9th, by which time a major salvage operation had swung into action.

peak water ballast tank and Nos. 3 and 4 port tanks. The only good news was that there was no evidence of oil spill — although the MSA declared a Tier Three response to the casualty, which enabled it to co-ordinate all appropriate agencies if a high-level response was required. After 10 days the ship was successfully refloated.

The *Tai Ping* incident had serious implications for South Port Ltd, which the MSA prosecuted (the first time it had prosecuted a port company). The fines imposed — $7500 to South Port for providing pilotage services and tug assistance in a manner likely to cause danger or risk, and $750 to the master of the tug *Hauroko* for failing to ensure that the ship had been navigated in accordance with MSA regulations — were minor compared to the damage to the port's reputation. Although everyone conceded the problems were created by the worst fog in 20 years, the port company had no contingency plan for such an event. The port pilot had asked the *Hauroko*'s master to act as his 'eyes' but the tug did not have its radar switched on and the tug master sent a deckhand outside to 'keep an eye out over the stern for *Tai Ping* to warn him if *Tai Ping* was about to overrun the *Hauroko*'. In any case, the *Hauroko* was lost and moved further to the north-east than it would have done had the tug master been able to see.[12] 'It was a case of the blind leading the blind.'[13]

15

'Rocks ahead! … Rocks astern!' — passengers in peril

We have become so accustomed to airline travel that it is easy to forget that, even today, ships still move hundreds of thousands of people around our shores. The Interislander and Strait Shipping services link the two islands; Auckland, Wellington and Lyttelton's harbour ferry services have revived; tourist catamarans and launches tout for business in tourist centres; and every summer a fleet of ever-bigger cruise ships descends on New Zealand. Sea travel seems so planned, so safe. Big steel-hulled ships equipped with sophisticated satellite navigation equipment should be safe from danger. Yet, as the wrecks of the *Wahine* in 1968 and the *Mikhail Lermontov* in 1986 showed, even the biggest, best-maintained ships can fall victim to the forces of nature or to human error.

The *Wahine* disaster

In the early hours of 10 April 1968, able seaman Alvyn Finlayson was straining to see anything through the murk from the port side of the ship's bridge. The Union Company's sleek new passenger ferry *Wahine* was entering Wellington Harbour under appalling conditions — torrential rain and gale force winds were toppling power lines and stripping roofs off houses in the capital they could barely glimpse. Finlayson's captain was just as confused. The *Wahine*'s engines were going, but the ship's violent pitching and rolling made it impossible to estimate either speed or direction. Then at 0640 the chief officer saw a light on the starboard bow, then one on the port bow. It was grim news. Seconds later he saw rocks to starboard. 'Rocks ahead!' he cried. Then someone shouted, 'Rocks astern!' Captain Hector Gordon Robertson raced for the starboard wing of the bridge and froze when he saw rocks on the starboard bow: 'Then I saw them astern and there was no way of getting out of it.' The company's flagship was about to hit. Robertson recalled his ship being 'picked up bodily and thrown on to the reef'. Strangely, he did not feel the initial impact, although he would always remember feeling the *Wahine*'s hull bouncing up and down on the rocks. 'It was the end of a ghastly half-hour of

OPPOSITE The *Wahine* capsizes in appalling conditions, rope ladders trailing down her sides.
Ref: PAColl-7796-85, Alexander Turnbull Library, Wellington, NZ

ABOVE Sleek, modern and magnificent, the new *Wahine* flies through its paces during trials in Scotland in 1966.
Gift of Wellington Museums Trust, New Zealand Maritime Museum (2012.0.10213)

OPPOSITE, TOP The *Wahine* lists heavily to starboard near Steeple Rock. The degree of list meant that only the lifeboats on this side of the ship could be launched.
Ref: 35mm-01149-28-F, Alexander Turnbull Library, Wellington, NZ

OPPOSITE, BELOW The rail ferry *Aramoana* reached the scene just before 1400 hours. Thirty minutes later, when this picture was taken, the *Wahine* had rolled over on to its starboard side and the sea was a mass of swimmers, life rafts, boats and flotsam. Horrified observers watched as steam gushed up as cold seawater rushed into the doomed ship's boilers.
Morris Hill, 10 April 1968. Ref: 1992.2968.6, Wellington Museums Trust Collections

commanding a blind ship in the most hazardous position imaginable.'

It was also the end of a magnificent ship with a promising future. How different it had seemed less than two years earlier. 'Sleek New *Wahine* is World's Biggest Roll-On, Roll-Off Ship,' the Union Company had boasted in June 1966 as its new Lyttelton Wellington ferry sailed up Wellington Harbour. For more than 70 years swift 'steamer expresses' (as it pompously called them) had come and gone so punctually that people said they could set their watches by them.[1] Everyone knew their names: *Rangatira, Hinemoa, Maori* and *Wahine*.

Everyone who saw the ship on that first day predicted that the *Wahine* would have a long and quiet career. At 8944 GRT it was one of the world's largest roll-on, roll-off passenger ferries. And stylish! The raked bow and masts, the fully enclosed wheelhouse and the prominent funnel blended old and new handsomely. The *Wahine*'s powerful turbo-electric engines gave a speedy 21 knots, with more in reserve if needed. The ship had the very latest radar and echo sounders.

But the *Wahine* was a 'bad luck ship'. Its difficulties had started in the shipyard, where labour disputes and the financial problems of the shipbuilder had delayed construction by almost a year. Mechanical 'teething problems' and a potentially fatal accident at Lyttelton, when a freak gust of wind carried the ship away from the wharf while people were still on the gangway, tarnished the ship's reputation.[2] Then, in April 1968, came the disaster that left such an indelible mark on our memories.

It began in the faraway Coral Sea on 5–6 April when tropical Cyclone Giselle sprang to life. In the next few days, the storm swept down into the Tasman Sea, gathering intensity. It passed over the North Island during 9 April, its winds peaking at more than 100 knots off the Wairarapa coast. Unfortunately, it then encountered a vigorous cold front in Cook Strait, slowing down and intensifying. Wellington reeled under the impact; hurricane-force winds ripped off roofs, sank small craft, and stretched the city's emergency services.

These were the appalling conditions into which the *Wahine* sailed on the morning of 10 April. The ferries had punched their way through some violent storms before, and at first Captain Robertson was not unusually concerned. The meteorologists said the storm was abating; with its radar and powerful engines, the *Wahine* should have been able to get its 610 passengers, 123 crew (and two stowaways, seamen 'ringbolting' their way to Wellington) into harbour safely.

But the storm swung south further than predicted. The *Wahine* was nosing into harbour early in the morning when, in deteriorating

'Rocks ahead!... Rocks astern!' **passengers in peril**

visibility, the radar failed just as the ship entered the narrow and rocky entrance. Several strong gusts of wind then caused it to sheer off course. Next, a massive sea struck the vessel on the port side, sending Robertson and his chief officer flying the length of the wheelhouse. For 30 minutes they struggled to regain control and head the ship out into Cook Strait, but to no avail. At 6.40 the *Wahine*'s luck ran out when the lookouts called out those fatal, fateful cries: 'Rocks ahead!', 'Rocks astern!'

The *Wahine* struck Outer Rock, the southernmost part of Barrett Reef, tearing open its bottom and wrecking the engines. Robertson closed all watertight doors and dropped anchors but the helpless ship drifted up harbour completely out of control. Chief Officer Rodney Luly and bosun George Hampson crawled on their stomachs along the deck to release the anchors, which offered some relief by holding the *Wahine*'s bow towards the oncoming waves and wind before they began to drag.

The port's tug *Tapuhi* headed out under the command of Captain Athol Olsen, assisted by pilot Captain Cyril Sword. They almost needed rescuing themselves: three times the over-100-knot winds blew this underpowered relic around helplessly. After sheltering briefly in Worser Bay, the old tug succeeded in attaching a towline to the ferry about 1120, but less than 20 minutes later the 102-mm wire parted. In danger itself, the *Tapuhi* wheezed for shelter while crewmen planned a way of towing the ferry into Worser Bay. In the meantime, the *Wahine* drifted closer to the shore and to total destruction.

By about noon the *Wahine* had drifted close to Steeple Rock, off Seatoun, where the last hours of the drama were acted out. For passengers the first real warning of danger had come six hours earlier

Survivors coming ashore at Seatoun in one of the ship's boats.
Ref: EP/1968/1576-F, Alexander Turnbull Library, Wellington, NZ

> Some passengers jumped into the sea, others simply fell. Those who made it into the starboard boats were the lucky ones. Even so, with the wind shrieking past and the rain pelting down, they went through hell before finally hitting the beach.

when some of them had felt the ship graunch across the reef. Reassured by ships' officers that they were in no danger, they mustered on B deck 20 minutes later. For the rest of the morning they sat around in their lifejackets, anxious but not yet frightened. The ship had started listing to starboard about 1010 but an extemporised lunch off Seatoun took their minds off it.

An hour later though, at 1300, the *Wahine* started to list badly and to wallow. The order to abandon ship came barely 10 minutes later. Suddenly shocked and dazed passengers found themselves climbing into lifeboats or scrambling over rope ladders down the side of the ship and into the surging waters. Leaving a ship — even one as obviously doomed as the *Wahine* — can be terrifying, especially if you are elderly, as many of the ferry's passengers were. Many people did not even know which was the starboard side.

The ship's list increased dramatically, throwing many off their feet and making it impossible to launch the port lifeboats. Some passengers jumped into the sea, others simply fell. Those who made it into the starboard boats were the lucky ones. Even so, with the wind shrieking past and the rain pelting down, they went through hell before finally hitting the beach. Less fortunate were those who ended up in the frigid water. Some swam to Seatoun but for most of those swept towards the rocky, inhospitable coastline on the eastern side of the harbour, small craft offered the only real hope of survival. These eastern shores, beautifully bleak on a sunny day, are inhospitable in winter. Virtually unpopulated, rugged and rocky, they accounted for most of the *Wahine* casualties, as exhausted swimmers were dashed against rocks.

A rescue response was delayed, according to the police history, by Robertson's delaying reporting the real degree of danger 'until the last moment'. Suddenly, emergency services, already busy responding to shore damage calls, had to switch attention to the harbour. Chief Inspector George Twentyman, who had worked on the 1953 Tangiwai railway disaster, co-ordinated rescue activities; eventually 371 of the 629 police members in the region would be involved.[3]

Rescuers were frustrated by the sea conditions. Just before 1400 the ferry *Aramoana* arrived on the scene. It gave a psychological boost to swimmers, but its great size meant that it had to keep away from them, launching boats and trying to act as a breakwater against the worst of the waves. The smaller craft — *The Portland,* HMNZS *Manga* and private craft — did most of the actual rescue work. Their crews worked frantically, since the water was bitterly cold, leaching vital warmth from swimmers' bodies. The first lifeboat was launched at 1330 amid great confusion; by mid-afternoon the first survivors were coming ashore at Seatoun and volunteers were being rounded up to bring blankets, clothing, food and tea. People from all over the city turned up to help. By the middle of the afternoon, the Wellington railway station — the marshalling station for the frozen, drenched and battered survivors not injured badly enough to be hospitalised — resembled a battlefield.

Early newspaper reports spoke of the possibility of 140 dead or missing; the final death toll was a still mind-numbing 51. Many were elderly people, swept across the harbour to the rocky eastern shore where most of the bodies were recovered. New Zealanders were stunned. Television crews recorded the shipwreck in a way never before possible, and tributes flooded in from all over the country. For years afterwards the pictures of the graceful, doomed ship sliding over onto its side would be a stock shot of documentaries and retrospectives.

The court of inquiry cleared the senior officers, the Union Company and the Wellington Harbour Board of all major charges of wrongdoing. It found that the main cause of the *Wahine*'s loss was that the ship — struck by one of the worst storms in New Zealand history — sheered off course in zero visibility, went out of control, and struck Barrett Reef, sustaining serious underwater damage. The immediate cause of capsize was free surface water on the vehicle deck. The court also stated that loss of life would have been much greater had the ship been abandoned off the entrance. While criticising Captain Robertson for not warning shore authorities of the damage sustained by his ship early enough, the court praised his seafaring skills once he found himself in trouble. It also said that it could not criticise his decision to enter harbour; the storm had changed direction and character, so much so that the assessment given the ship at 0500 was no longer valid an hour later when he was entering the harbour. It also made a number of recommendations about ship design (although it could not fault the *Wahine*'s basic design — it was superbly built and well equipped) and

As if to taunt, the weather fined up just days later. This aerial view shows how close the *Wahine* was to Seatoun when it capsized.
Ref: EP/1968/1571/25-F, Alexander Turnbull Library, Wellington, NZ

One of the *Wahine*'s masts has been preserved at Frank Kitts Park as a memorial to the ship and to the people who died on 10 April 1968.
Gavin McLean Collection

recommended that the harbour board acquire modern salvage tugs — the 'big reds' that still dominate port activities.

It was all over. Relatives buried their dead, the harbour board ordered its tugs, shipping companies sharpened up shipboard safety procedures, and the Union Company ordered a replacement ship, the *Rangatira,* which maintained the increasingly uneconomic run until 1976.

The *Wahine* was not left to rest in peace because the harbour board insisted on removing the wreckage. At first, optimistic salvors talked of refloating the ship intact by pumping it full of polyurethane foam. But as each successive storm bashed and battered the exposed wreck, scrapping on-site became the only option. Those storms almost broke the salvors — United Salvage Proprietary Ltd — as operations dragged on. Its chartered coaster *Holmpark,* assisted by the floating crane *Hikitia,* became fixtures off Seatoun for years as the company and its successors struggled to complete an increasingly burdensome contract. Today little remains to mark the dramatic events off Seatoun 51 years ago. There are a propeller, anchor chain and a plaque at 'Wahine Park', Seatoun; lesser memorials at Palmer Head and Eastbourne; and the Museum of Wellington City and Sea has a large-scale model of the ship in its death throes, complete with lifeboats and struggling swimmers as part of a moving *Wahine* multimedia exhibition. In 1990 Lambton Harbour Management Ltd erected one of the masts, gifted to the museum by a prominent entrepreneur, in a place of honour in the newly refurbished Frank Kitts Park. The other has been erected in the Hutt Valley.

The *Mikhail Lermontov* mystery

The Soviet liner *Mikhail Lermontov* became New Zealand's biggest shipwreck in February 1986 when it foundered in Port Gore in the Marlborough Sounds after striking rocks.[4] The stylish, white-hulled ship was no stranger to New Zealand waters. The last of the five *Ivan Franko* class built at Wismar in East Germany in 1972, the *Mikhail Lermontov* had made many cruise voyages to New Zealand. It was big. Some 19,872 GRT and 176 metres long, when built the *Lermontov* could accommodate 750 one-class passengers and a crew of 300. Twin 21,000 shp (shaft horsepower) Sulzer diesels gave a service speed of 20 knots. Just four years before its final voyage, a Bremerhaven shipyard had refurbished the ship to offer Western-style luxury for 550 passengers. In its new guise the *Mikhail Lermontov* measured 20,352 tons.

In February 1986 it was in the middle of a four-and-a-half-month cruising season for CTC Lines of Sydney. Nine 11-day cruises were planned, together with a 41-day Asian voyage. The ship had visited Auckland and Tauranga before arriving at Wellington on the morning of 15 February. After giving passengers a day exploring the capital, the ship sailed around midnight, bound for Picton. There it offered passengers a short shore break before starting the return journey to Sydney via Milford Sound.

The ship left Picton at 1510. Marlborough Harbour Board harbourmaster Captain Don Jamison (who also held a pilot's licence for Milford Sound) was on the bridge, pointing out places of historic interest as he took the ship up Queen Charlotte Sound. At 1635 the master, Captain Vladislav Vorobyov, retired to his cabin, leaving instructions to be called back to the bridge when the *Mikhail Lermontov* reached Ship Cove. He was notified at 1700 but he did not reappear, effectively leaving Jamison in charge.

As the ship neared the head of the sound, it was steering a course of 0400T, which should have taken it well clear of trouble. When the passage between Cape Jackson and its offshore lighthouse opened up, Jamison ordered the helmsman to 'port 100'. On hearing this, second mate Sergey Gusev warned chief navigator Sergey Stephanischlev that Jamison's course was taking the ship into danger. Stephanischlev queried Jamison but was told that the pilot was merely taking the ship closer to Cape Jackson to give passengers a better view. When the clearly worried Stephanischlev again questioned the wisdom of Jamison's move at 1734, he was told that the pilot intended to take the ship through the passage. Stephanischlev neither sent for the master nor overruled Jamison.

By now it was probably too late. The *Mikhail Lermontov* swung from 0400 through north to approximately 300 degrees and headed through the narrow passage, slightly nearer to the shore than to the light tower. At 1737, while steaming at

Unfazed by the drizzly weather, many of *Mikhail Lermontov*'s 408 passengers took the opportunity to explore the town of Picton and the nearby vineyards of Blenheim. Most were elderly Australians, enjoying what tour brochures advertised as 'the cruise of a lifetime' on a ship that was commanding presence wherever she berthed.
Ref: EP/1986/0740/34-F, Alexander Turnbull Library, Wellington, NZ

This dramatic photograph — taken by fisherman M.H. Harris — shows the *Mikhail Lermontov* beginning to founder late on the night of 16 February.
Ref: EP/1989/1713/4, Alexander Turnbull Library, Wellington, NZ

15 knots, the *Lermontov* grounded on its starboard side, just forward of amidships. The ship kept moving but listed to starboard as water rushed in. Vorobyov returned to the bridge and, realising the gravity of the situation, turned his ship into Port Gore, intending to beach it before it sank.

He almost made it. The ship's designers had allowed for the flooding of two watertight compartments but four were already awash and the ship's generating room was damaged. By the time Vorobyov beached in Port Gore an hour later, his ship had lost almost all power and was down by the bows. It drifted onto a sandbank and waited for help. Unfortunately, the incoming tide started to push the ship off the sandbank and out into deeper water. Some suggested that he should have dropped the anchors to keep the ship on the sandbank, but Vorobyov said that he feared that might harm or panic people in the water and that he would not be able to lift them again when the ship's power failed.[5] The *Mikhail Lermontov* sank at 2050. Witnesses recalled the terrifying spectacle as huge bubbles belched from the sea, and loose items from the ship shot to the surface, leapt into the air and then noisily smacked down on the surface of the water.

Ships converged on the scene quickly. Captain Vorobyov had refused offers of assistance, but fortunately the masters of the LPG tanker *Tarihiko* — which had been sheltering off D'Urville Island — and the rail ferry *Arahura* had decided to stick close. Other rescuers were the cement carriers *Golden Bay* and *Milburn Carrier*, and the patrol boats *Taupo* and *Wakakura*. Also dispatched from

Wellington were the tugs *Kupe* and *Toia* (both of which were turned back) and the police launch *Lady Elizabeth II*.

Although heavy rain and moderate seas made the night-time rescue difficult, the little flotilla picked up everyone, except for refrigeration engineer Pavel Zaglyadimov, who is believed to have been killed when the *Lermontov* struck the rocks. The *Arahura* and *Tarihiko* took passengers and crew to Wellington next morning. Considering the remoteness of the accident, the weather and the age of many passengers, they got off lightly: only 11 passengers suffered minor injuries, and by that evening most were flying home.

Geo H. Scales, CTC's Wellington subagents, had been kept busy, finding divers, answering phones and even drying piles of waterlogged banknotes in clothes dryers.[6]

Then the real trouble started. The Marine Section of the Ministry of Transport held a preliminary inquiry under Captain Steve Ponsford. In addition to praising the crew for helping the passengers and declaring the ship's lifesaving equipment adequate, it found that Jamison, who was operating the ship just outside the limits of the Marlborough Harbour Board pilotage area, had taken the ship through a narrow channel which he knew was too shallow for a vessel that size.

Why he did so remains a mystery. At that point Transport Minister Richard Prebble announced that, since the cause of the accident had been established, no formal inquiry would be held. As we have seen, this was not unusual — there was no inquiry into the loss of the *Niagara*, for example, because the cause was found quickly and simply. But people wanted to know 'why' as well as 'what' and 'when' — especially angry passengers, and some mariners who felt that unanswered

All that remained ... a launch tows a brace of boats back to port.
Ref: EP/1986/0770/18A-F, Alexander Turnbull Library, Wellington, NZ

> **Witnesses recalled the terrifying spectacle as huge bubbles belched from the sea, and loose items from the ship shot to the surface, leapt into the air and then noisily smacked down on the surface of the water.**

questions might blacken the reputation of New Zealand pilots. But Minister of Transport Richard Prebble refused and an MOT decision to release the full transcript of the preliminary hearing was blocked by legal action from Jamison.

The *Lermontov* remains controversial, generating many books, TV programmes and newspaper articles. 'This has to be one of the most clear-cut cases ever,' Prebble commented in 2006. 'The pilot, Captain Don Jamison, never disputed that he made the decision to take that route and it was wrong and that was why the ship sank ... But why he decided to guide the ship over a passage that he actually knew was too shallow, I don't think he'll ever be able to answer'.[7]

The Soviets held their own inquiry which also blamed Jamison's decision to take the ship through a dangerous channel too shallow for a vessel of the *Mikhail Lermontov*'s size. It censured Captain Vorobyov (who was transferred to a shore job), Stephanischlev (sentenced to four years' imprisonment) and Gusev (who had his certificate suspended for two years). Salvage was considered uneconomic and since the wreck lies far from shipping channels, it was left undisturbed. Between March and August 1986 teams working from the *Little Mermaid* recovered the oil (about 1300 tonnes) from the *Lermontov* and secured the wreck site.

The ship lies in Port Gore in about 37 meters of water and, despite the dangers of the site, is dived on regularly by recreational divers, who have removed many fittings and contents. But such activities come at a price — since the sinking, three divers have drowned there.[8] Pete Mesley, who dived on the wreck in 1999, wrote about the dangers inherent in such large and recent wrecks. 'Swimming along the prom deck, our first port of call was the Rainbow Cinema. All the chairs were still in their rows, slowly rotting away with time. The ceiling had started caving in and the stage was collapsing, so utmost care had to be taken'.[9] Nevertheless, thousands have dived on what has been described as the world's third-largest diveable wreck.[10]

'Ferry seconds away from disaster'

Fortunately, no other large passenger ships have sunk on the New Zealand coast since the *Wahine* and the *Mikhail Lermontov*. But ferry traffic has increased and, along with that 'marine occurrences', as officialdom calls them. In 2004 and 2005 alone, the passenger catamaran *Tiger III* was wrecked at Cape Brett; the Gulf ferry *Quickcat* was involved in a collision that had fatal consequences for a passenger on another craft; passengers on the ferry *Black Cat* were injured after an engine malfunction caused the ship to hit a rock in Akaroa Harbour; and the coastal cruise vessel *Milford Mariner* grounded briefly in Milford Sound. Cheap airfares have done little to reduce passenger traffic on Cook Strait. The 1990s brought stiff competition to the privatised Railways (Tranz Rail from 1995 and Toll New Zealand from the early 2000s) as high-speed ferries challenged the traditional ships before concern about wash-damage to shore lines torpedoed the 'vomit comets' in the early 2000s. In

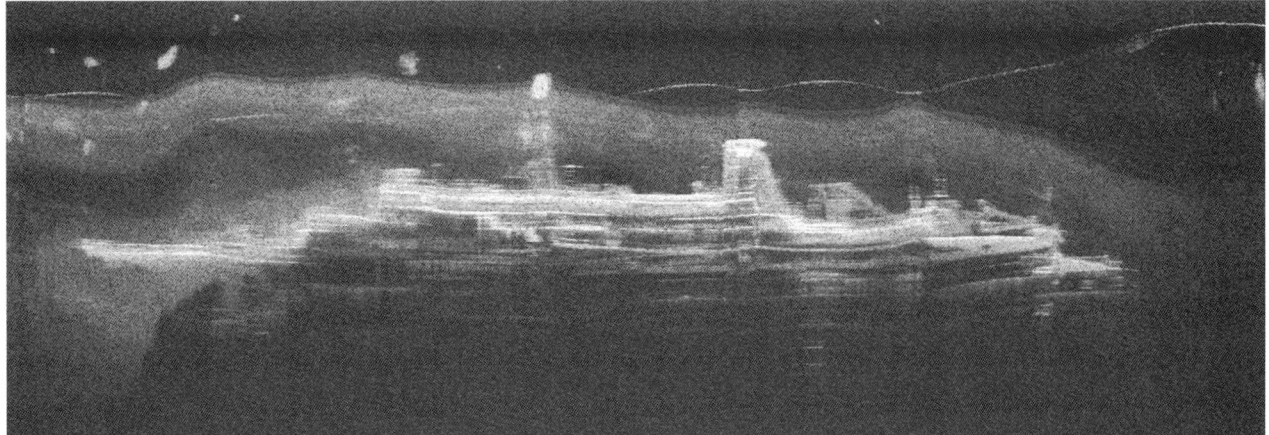

The ghost-like sonar outline of the *Mikhail Lermontov* lying in its watery grave in the Marlborough Sounds. Produced using side-scan sonar, the funnel, main davits, bridge, and various decks are clearly visible. The murkiness of the bow owes principally to its being partly buried in the mud.
K. Grange, NIWA

the days of government ownership there had been a few incidents, but the ferries were better known for school holidays strikes than for accidents. That changed in the 1990s with crew size reductions, the introduction of integrated ratings (multi-skilling, with deck and engineering crew being expected to perform each other's tasks), and inter-company competition.

In May 2006 readers of the *Dominion Post* were treated to a dramatic (and somewhat unrealistic) computer generated image below a screaming headline 'Cook Strait Ferry *Aratere* Nearly Capsized'. The 'Aradago', as industry critics labelled the 1999 Spanish-built ship, had been no stranger to bad press: the teething problems to be expected from nearly every new ship had seemed to dog this ship year after year. 'Cook Strait Ferry Seconds Away from Disaster,' the *Dominion Post* reported in 2004. On the night of 29 September 2004, the ship, with 268 passengers aboard and under automatic navigation, came 'within thirty seconds of potential disaster' while entering Tory Channel.[11]

Just days later it was back in the news when it was reported that the ferry had 'cut off' the freight ferry *Kent* on 1 October as it entered Wellington Harbour. The MSA moved in to audit the line's navigational practices and systems. Then on 19 October the 'Aradago' suffered another complete power failure between Allports Island and Dieffenbach Point in the sounds while en route to Wellington. It anchored and returned to Picton once power had been restored, transferring passengers to the *Arahura*.[12]

But the May sailing reported by the *Dominion Post* had been worse than these incidents. It was rough in the strait — reports mentioned peak average southerlies of 49 knots, gusting up to 63 knots. Passengers reported the *Aratere* twice heeling over on its starboard side by about 50 degrees, sending furniture sliding across cabins and throwing vehicles around on the lower decks, damaging about 20 vehicles and toppling six rail wagons. Television news screened dramatic amateur video of the wildly thrashing seas, while the *Dominion Post* reported a maritime expert as saying that 'an Interislander ferry came "extremely" close to capsizing after its captain made "foolhardy" decisions crossing Cook Strait in rough seas'.[13]

Maritime unions demanded the unthinkable: a say in determining whether ships should sail in bad weather. Maritime New Zealand stepped in and suspended the master's certificate pending the completion of its investigations. He appealed, then withdrew his appeal, accepting the required additional training and being reinstated to command before the year was out.

16

From horror to heritage

Our Victorian ancestors, knowing what it was like to be 'in peril on the sea', may not have liked being reminded of shipwrecks. But it would have been hard not to be aware of them, for shipwrecks often leave their mark long after the sea washes away the last physical traces. Numerous wrecks would become rooted in local memory through place names — places such as Nelson's much-photographed Fifeshire Rock, named for the immigrant ship that struck on 27 February 1842 after bringing the first party of settlers to Nelson or Timaru's Benvenue Cliffs, or Delaware Bay, Nelson, named after the brigantine wrecked there in early September 1863. Maps and charts similarly immortalise old blunders.

Culturally, maritime mishaps and tragedies have also been significant elements of local lore in some regions — the place of the wrecking of the *Nora*, *Manuka*, and *Surat* off the Catlins coast, for example, seeing the Catlins Area School houses named after the wrecks. Wrecks of iron and steel ships can survive for a very long time if they are not exposed to regular wave action. In time some became local landmarks or picnic destinations. On Otaki Beach the iron-hulled sailing ship *Hydrabad,* wrecked so high above the normal wave zone that it can still be seen, was visited by generations of children out for a day at the beach. Not far away, a mast from another sailing ship, the *City of Auckland,* protruded from the sand for 50 years after the ship went ashore in October 1878. Since 1987, when its largely intact hull was recovered from the beach of Fitzroy Bay, the remains of the small coastal steamer *Paiaka,* wrecked in FitzRoy Bay on 9 July 1906, have been on display to beach walkers.

Indeed, 'dark heritage' such as shipwrecks have long held a popular fascination for people, and here New Zealand has been no different. Unfolding stories of missing vessels, tragedy (and at times human endurance and heroism) captivated newspaper audiences — the story of the *Wairarapa* and her heroic stewardesses in chapter 8 being but one example. Early newspaper accounts of the experiences of the crew of the *Grafton* attracted such interest that when the *Flying Scud* returned to the islands and found the skeleton of James Mahoney (one of the castaways

OPPOSITE The *Hydrabad* is one of the country's most visited wrecks. The 1350-ton ship ran ashore in 1878 and has been visited by generations of people. This photograph was taken about 30 years after the wreck.
Ref: 1/2-075568-F, Alexander Turnbull Library, Wellington, NZ

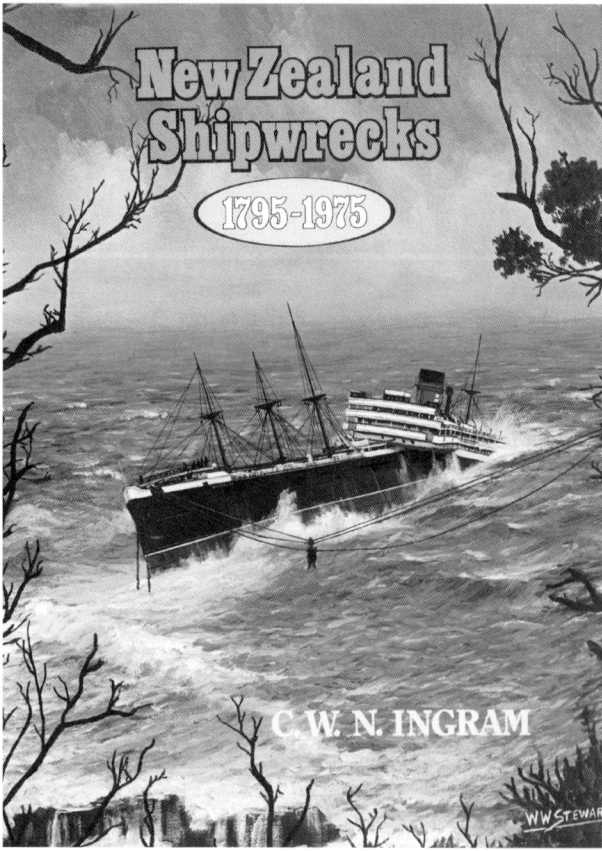

ABOVE Many editions of C.W.N. Ingram's *New Zealand Shipwrecks* were published, attesting to the popularity of the subject.
Gavin McLean Collection

TOP Nelson's Fifeshire Rock, from an old postcard. It is one of many natural coastal features named after a shipwreck.
Gavin McLean Collection

from the *Invercauld,* which had wrecked at the other end of Auckland Islands at the same time as the *Grafton*), an official mission was dispatched to search for more survivors.

Shipwrecks were an ever-present theme of New Zealand publishing from the early twentieth century, with stories of wrecks and survivors being common threads of the waves of regional and community commemorative histories that were published from the 1910s. Here wrecks were typically illustrative of either the challenges and threats facing the stoic pioneering generations, or as part of early collections of historical reminiscences such as Louis Ward's *Early Wellington* (1929), or Colin MacDonald's *Pages from the Past* (1933). The place of wrecks in New Zealand's national consciousness was only further nailed home with the 1940 Centennial publications, particularly those earliest incidents such as the *Boyd* and 'Harriett affair', which were presented as having some sort of formative influence on the nature of New Zealand colonisation.[1]

The appearance of dedicated volumes on shipwrecks also coincided with this period, starting with the 1936 publication by two Dunedin fire fighters, Ingram and Wheatley, of the first edition of their New Zealand reference book, *New Zealand Shipwrecks and Maritime Disasters*. A.H. & A.W. Reed published many later editions of *New Zealand Shipwrecks* for three decades and put out a string of books on famous wrecks: Joan McIntosh on the *Tararua* (1970), Keith Eunson on the *General Grant* (1974), and Max Lambert and Jim Hartley on the *Wahine*. Clif Reed himself also got in on the game, writing *The Wreck of the Osprey* in 1937 as a Raupo School reader. Other publishers followed suit, and a little later, books on shipwrecks by divers such as Wade Doak, Steve Locker-Lampson and Ian Francis and the entrepreneurial activities of Kelly Tarlton brought a new dimension by bringing wrecks 'alive' in photographs and words that showed what could be seen beneath the waves.

Fascination with the material heritage of

shipwrecks was a direct extension of interest in published accounts, though in many respects also preceded publication. Fascination with the *Grafton*, for example, never waned, and by the late-1880s the material remains of the vessel and castaways' plight were being sought by museums and history buffs alike. William Dougall's published account of his visit to Epigwaitt site of the castaway's encampment in 1888, for example, saw him return not just with a number of photographs of the Grafton wreck and Epigwaitt hut site, but artefacts — a number of which were deposited with Dominion Museum. *Grafton* objects were similarly added to the Canterbury Museum collection following 1901 and 1907 visits to the Auckland Islands.

Such interest was equally a formative influence on early heritage legislation in New

EDWIN FOX

One of the more specialist maritime museums is the old East Indiaman *Edwin Fox*. The ship was built in 1853 at Sulkeah, Bengal Province, for the Honourable East India Company and had an interesting career, serving as a transport in the Crimea War, a convict ship to Australia and an immigrant ship to New Zealand. Its seagoing days came to an end in 1885 when Shaw Savill & Albion converted it into a freezing storage hulk. The *Edwin Fox* served at Port Chalmers, Lyttelton, Gisborne and Bluff before coming to Picton in 1897, never to leave.

When the need for a floating freezer passed, the *Edwin Fox* served as a landing stage and a coal hulk before being abandoned in a backwater. Every now and then someone would write an article on the *Edwin Fox* or suggest its restoration, stirring up a brief flurry of interest before things died away again, leaving the old derelict to the attentions of vandals and decay.

There was a brief tussle between Wellington and Picton in the 1990s when a group of businessmen planned to take the *Edwin Fox* to the capital, but Marlborough rallied, and the old hulk was finally given its own roofed-over graving dock on the Picton waterfront. Visitors can browse a well-stocked museum of artefacts and inspect the ship itself.

TOP & ABOVE Visitors can inspect the remains of the *Edwin Fox* in its roofed-over graving dock at Picton.
Gavin McLean Collection

'PLIMMER'S ARK' — THE WRECK WITH ALMOST AS MANY LIVES AS A CAT

No wreck is more expensively conserved and interpreted than the *Inconstant*. The ship, just a year old, hit rocks while entering Wellington on 3 October 1849. Refloated badly damaged, it was beached at Te Aro and bought by merchant John Plimmer for use as a floating warehouse.

The appearance of the ungainly structural additions to the *Inconstant* gained it the nickname of 'Plimmer's Ark'. In 1883 they were demolished, and the hull's frames were cut down to make way for a new building on what was now reclaimed land. But it was not forgotten and in 1899, when the site was being excavated for another new building, Plimmer, by now the self-styled 'Father of Wellington', posed with his dog Fritz on the ship's temporarily exposed remains.

The 'Ark' made another comeback in the 1990s during the conversion of the former Bank of New Zealand complex into the Old Bank Arcade. Part of the ship can be seen in the building's basement, together with a display of items from the ship. Other parts are in the Shed 6 Gallery on Queens Wharf, where visitors can see them being conserved.

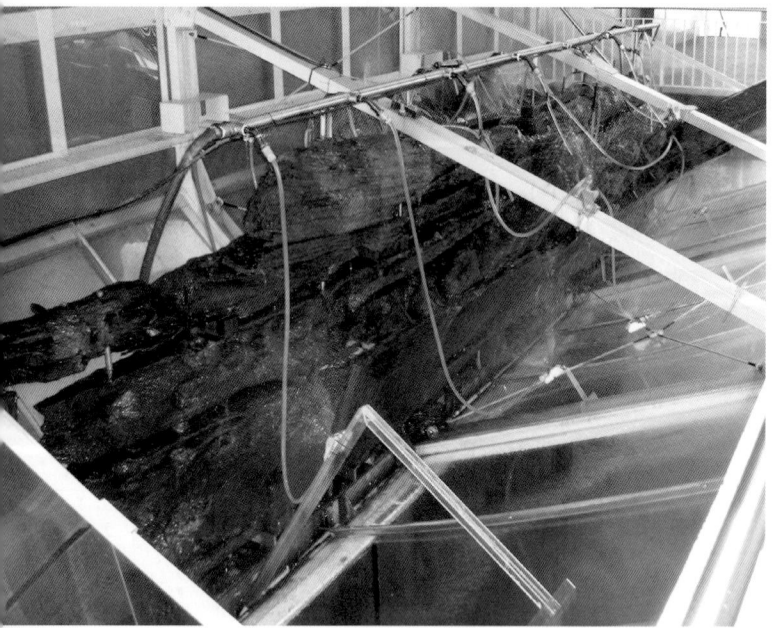

ABOVE A section of the *Inconstant*, on display in the Queen's Wharf Gallery. The sprays are part of the conservation treatment.
Gavin McLean Collection

TOP Workmen pose on the remains of the *Inconstant*, exposed during building excavation work in 1899.
Ref: 1/1-002704-F, Alexander Turnbull Library, Wellington, NZ

Zealand, with collector and tourist interest in the remnants of old and wrecked waka being a decisive factor in the drafting of New Zealand's earliest heritage legislation — the 1901 Maori Antiquities Act.[2] That same year the merchant and bibliophile Alexander Turnbull made a cruise to the Marlborough Sounds in his yacht *Iorangi* in search of the places at Ship Cove where Cook had made repairs to *Endeavour* from an earlier grounding; and Grass Cove where Captain Furneaux and two of his crew had reputedly been massacred by local Maori during Cook's second voyage. The following year, armed with a packet of photographs from Turnbull, the photographer and explorer Russell Duncan, arriving in Queen Charlotte Sound, opened the packet of photographs to find they were actually at Grass Cove, and sitting but metres from where the men had reputedly been killed. The feeling at discovering this, Duncan noted, 'was rather awesome, notwithstanding the 128 years that had elapsed, the occurrence and details being vividly before our minds'.[3]

But it was after World War II that the rise of interest in New Zealand's maritime history really took off. In part this was simply part of the remarkable post-war rise of concern for the preservation of cultural heritage that in New Zealand culminated in the establishment of the Historic Places Trust in 1954, however, wreck heritage was also more specifically associated with the rise of interest in 'industrial heritage' from the 1960s. There was, of course, also a technological element to this, with the invention of the aqualung by Jacques Cousteau and Emile Gagnan in the 1940s simply giving underwater explorers more freedom than the heavy, old, tethered diving suits previously used by harbour boards and professional salvors. Although it was not always easy to get the right equipment before the 1960s, scuba gear made underwater equipment affordable for weekend explorers — indeed, as a result of this, underwater archaeology in New Zealand really has its beginnings in the 1950s and early 1960s. Cousteau's books and televised exploits only further boosted such interest.

Kelly Tarlton did more than any New Zealander to improve public awareness of wrecks. An enthusiastic, self-taught diver, Tarlton dived on several significant shipwrecks. The coins — about 1.5 tonnes of mainly silver coins — and the four bronze propeller blades he recovered from the wreck of the *Elingamite* in the late 1960s, attracted considerable public interest, but it was Tarlton's discovery that the propellers were bent, that showed the scientific value of shipwrecks. The 1903 inquiry had been unable to determine whether the ship's screws were still turning when it hit, but this new evidence supported the master's evidence that they had hit rocks earlier, sustaining damage that proved fatal. Divers still blasted and looted, but now they could also be historical detectives and archaeologists.

After a dive in 1974 on the *Tasmania,* Tarlton researched the history of the ship and found that jewellery merchant Isadore Rothschild had lost a chest of jewellery when the ship sank off Mahia Peninsula in 1897. That sparked his interest and saw him organise several expeditions to the ship which lay heavily buried in sand in 40 metres of water, well offshore. By 1978 he had recovered over 300 pieces of jewellery but needed to establish ownership rights. Fortunately, Rothschild's nonagenarian daughter was still alive in Melbourne and sold the rights to the jewellery and the ship which she had inherited from her father who had once dreamed of raising the ship with 2000 rubber balloons.[4] Much of the horde from the *Tasmania* and the *Elingamite* were stolen in 2000 by an employee who never divulged their hiding place, saying it would be 'worth more than his life' to do so.[5] Despite exhaustive efforts to recover the loot — including investigators offering rewards, making pamphlet appeals through prisons, and searching the thief's childhood eeling spot in a Northland stream — none of the stolen treasure has been recovered.

Tarlton was also involved with two of the

country's earliest wrecks. The first was not a wreck but had almost been one. In 1769 de Surville's ship *St Jean Baptiste* had lost three anchors and narrowly escaped destruction while berthed in Doubtless Bay. An error in the transcription of an old chart had Tarlton foxed for some years, but after checking the original and allowing for other factors, Tarlton discovered all three anchors.

Tarlton's other early wreck was the *Endeavour* at Facile Harbour in Dusky Sound. Whalers had cut away the upper parts of the hull in the nineteenth century, but a considerable amount of the lower hull remains, buried in sand and silt. HMNZS *Tui* visited in 1963 and reported that the ship's outline could be seen clearly in the water. In 1984 Tarlton, using a proton magnetometer, located the two cannons lost overboard while being transported ashore — after first dislodging a conger eel.

The idea of a museum of shipwrecks had been in the back of Tarlton's mind for years. In 1968, fresh from his dives on the *Boyd* in Whangaroa Harbour, he set up a small museum in an old custom shed overlooking the harbour. As few visited this isolated place in winter, he decided to move to the Bay of Islands where tourist numbers were greater. In 1968 he bought an old sugar lighter, the *Tui*, built in 1890 for the Chelsea Sugar Refinery in Birkenhead. Working on a shoestring budget, and battling bureaucracy, Kelly and Rosemary Tarlton somehow ensured that the 'first visitors walked aboard and paid their 30 cents on 12 January 1970'.[6] A few years later Tarlton rigged the old lighter with three masts recovered from the barque *Endeavor II* which was returning from a re-enactment of Captain Cook's landing at Botany Bay in 1971 when it was caught in a storm off North Cape and wrecked. Tarlton bought the masts and rigging and stored them near the *Tui* until his finances permitted him to re-rig the *Tui* as a barque. Many of the artefacts had been collected by Tarlton himself and for many years the Rothschild jewels were the highlight of the Museum of Shipwrecks. Tarlton died in 1985; Rosemary Tarlton sold the vessel and collection in 2002.

E.V. Sale, one of Tarlton's biographers, called the work of the first-generation wreck hunters 'robbery without violence'. Photos of columns of water shooting up from wreck sites featured in those early books on wreck exploration.

The desire to pay for their hobby by picking up coins or selling bronze propellers to scrap dealers drove some divers to devastate wreck sites. As marine biologist Mike Bradstock lamented, until late last century many amateurs worked over wrecks in a haphazard, ad hoc way. As a result, significant artefacts were scattered throughout the country in private collections, seldom recorded, photographed or given proper preservation care. This was looting, for 'shipwreck sites more than most other archaeological sites encapsulate a moment in history when a wide cross-section of objects relating to a time become lost', Bradstock observed. 'For ships are the homes of their crews (and passengers, if any) and as such contain most of the impedimenta of everyday life with nothing saved because the loss is usually sudden and unforeseen.'[7]

The shift in mindset from shipwrecks as treasure troves or free scrap yards to sites requiring preservation and conservation was recognised by the 1975 Historic Places Act, which placed shipwrecks 100 years old or older under the same legal protection as terrestrial sites. Under

OPPOSITE, TOP Environmental concerns and labour costs have made ship-breaking impossible in New Zealand. Most old ships are now stripped and scuttled at sea. After the *San Domenico* was written off after being hit in a berthing accident at Wellington, it was scuttled in Cook Strait on 6 July 2004.
Phil Hoyland, courtesy of Michael Pryce

OPPOSITE, BELOW Most former navy vessels have been scuttled to form artificial reefs and diving attractions. The *Leander* Class frigate *Wellington* was brought back to the capital in 2006 by the Sink F69 campaign. Prepared first by the stripping of fittings and potential environmental contaminants, the top hamper was then removed, and the hull was fitted with sink holes. On 13 November the ship was towed to a point off the southern coast, watched by a large fleet of spectator boats, after which scuttling charges were detonated. Sunk, the *Wellington*'s keel sits at roughly 21 metres, making the wreck accessible by scuba divers using standard equipment.
Michael Pryce

this legislation (amended in 1993 to cover any wreck or underwater site of cultural significance containing evidence relating to pre-1900 events), written permission has to be obtained from Heritage New Zealand (formerly the Historic Places Trust) before removing artefacts. In 1983 an Auckland salvage operator was prosecuted for removing material from the *Taupo,* a steamer which foundered off Mayor Island in 1881 while being salvaged after sinking at Tauranga two years earlier. Local divers, who rediscovered the wreck in 1979 in 'remarkably good condition', were negotiating with officials to have the site declared an underwater reserve, when the man damaged the wreck.[8] Judge D.B. Wilson convicted him in a judgment 'of major importance to the Trust in its use of the legislation to protect archaeological sites', but he won an appeal to the High Court, which decided that it could not be proved that he knew the wreck was over 100 years old.

Central as the Historic Places Act is, this is in reality part of a three-frame legislative framework for underwater heritage, the others being the Protected Objects Act (1975) — which regulates the export of protected objects; and the Resource Management Act (1991) — which focuses more on wreck remains as archaeological sites, and requires local authorities to protect wreck sites from inappropriate use and development.

Despite this legislative framework providing the basis for the control of both submerged and terrestrial archaeology in New Zealand, maritime archaeologists argue that its application to underwater cultural resources has yet to be

Archaeologist Greg Walter (left) and Auckland Council Cultural Heritage Specialist Rob Brassey near the remains of The Daring soon after it's rediscovery in May 2018. Despite the wreck laying within the NZDF Kaipara Air Weapons Range, within days of its rediscovery it had been targeted by relic hunters. Six months later the 'Daring Rescue Team' set about the ambitious task of recovering the vessel intact from its sandy grave, and by early 2019 it was safely in Hobsonville where it has been receiving stability and preservation treatment while a permanent home is established.
New Zealand Defence Force

undertaken in a systematic or significant way.[9] Particular concern also rests with the scope of how wrecks and wreck sites are defined, such as the need for clearer consideration of Maori underwater heritage sites and sunken canoes, some of which date from as early as the mid-thirteenth century, however, protection for them was extended with the Heritage New Zealand Pouhere Taonga Act 2014.

Nonetheless, limitations to the legislative framework were central to the failure in 1998 of the Maritime Archaeological Association of New Zealand (MAANZ) efforts to save the recently rediscovered wreck of the small steamer *Aorere,* which went ashore in 1921. The wreck had been covered in a sand dune and further protected from fossickers by being inside an air force bombing range. In 1998, however, archaeologists returned to discover that a Whanganui man had chainsawed away part of the wreck, with the intention of displaying it in a private museum. The Trust was helpless to act because the wreck occurred after 1900.[10]

MAANZ was formed in 1989 to record and catalogue wrecks and other underwater artefacts and to promote public interest in and respect for maritime archaeology.[11] The organisation has conducted underwater surveys and conserved artefacts in its conservation laboratory in the steam crane *Hikitia* at Wellington. Even so, Heritage New Zealand estimates that only 200 of the country's more than 2500 wrecks have been accurately located.[12]

In recent years the practice of underwater archaeology in New Zealand has been furthered in three main ways: through the work of volunteer groups, the teaching of Australian Institute for Maritime Archeology (in conjunction with the British Nautical Archaeology Society) training courses, and through the registration of shipwrecks as historic places. The timing of this has been critical as coastal and underwater sites have become increasingly threatened by the plans of other users of the coastal areas, which has seen increasing pressure on sites through the construction of infrastructure such as marinas, port expansion, dredging, marine farming and mineral extraction.

Such developments also paralleled a renewed interest in New Zealand's maritime history that developed from the late-1980s. The Wellington Harbour Board's modest maritime gallery went through a number of transformations to end up as the Museum of Wellington City & Sea. Auckland built a brand-new New Zealand National Maritime Museum. Smaller maritime museums exist in places such as Lyttelton, Port Chalmers and Bluff, and even in museums that are not explicitly maritime, figureheads, lifebelts, signalling guns and brass portholes attest to the power of the sea.

There is nothing new about erecting memorials to people drowned in shipwrecks. A large monument has long graced the intersection of Timaru's Perth and Sophia Streets, and the New Zealand Historic Places Trust marked the site of the what was then called the 'Boyd Massacre' relatively early in its history. But the main wave of memorial construction has occurred relatively recently. In the South Otago region, three major shipwrecks have been commemorated with publications and/or the unveiling of plaques. In 1969 the Wairarapa Underwater Club decided to erect an anchor on the beach at Mangatoetoe from the barque *Ben Avon,* wrecked on the coast in 1903, as a memorial to that and to the other ships wrecked locally. Anchors are the most popular artefact for use in memorials. Wellington has masts, anchors and propellers from the *Wahine* and memorial plaques at five locations. Wellington even has a self-guided tour booklet for a walk through the Karori cemetery graves of victims from the *Penguin.*

Renewed interest in maritime history and a strengthened protective framework have also been timely as climate change and increasing coastal erosion have in recent years uncovered hitherto forgotten or lost wrecks, such as discovery of the wreck of the *Hippolas*, which struck Walker Rock in 1909 and was abandoned with no loss of life, or the 'mystery wreck discovered off Wellington's Owhiro Bay in mid-2018.[13] By the years end, shifting sands had also uncovered the surprisingly intact remains of *The Daring* on Muriwai Beach—the old schooner having been beached there in 1865 after its first-time skipper, Captain Phipps, mistook Manukau Heads for Waikato Heads and kept sailing further north.[14]

With the demise of the small local shipbreaking industry, scuttling old ships at sea to form artificial reefs has become common. The *Rainbow Warrior* was an early example, but since then the navy's old *Leander* class frigates have been scuttled, with the sinking of the *Wellington* in 2005 off the city's south coast gaining front-page publicity. As we will see in chapter 17, the MV *Rena*, which grounded on the Astrolabe Reef off the Tauranga coastline in 2011, has become yet another popular site for a weekend dive.

Whether the subject is a scuttling or a historic wreck, increasingly, people will encounter sunken ships in their natural environment. Weekend scuba divers and tourists patronising the country's dive companies are gaining deeper understanding of our underwater archaeology. There is little fear of the resource drying up. Time and tide erode everything, but we can be sure that the sea will continue to exact a toll on the unwary, the inebriated and the unlucky.

A diver exploring the wreck of HMNZS *Canterbury*. Decommissioned in 2005, the *Canterbury* was scuttled in Deep Water Cove near the tip of Cape Brett in 2007. Constituting part of a dive trail that also includes the *Waikato* and *Tui* off Tutukaka, and the *Rainbow Warrior* at Matauri Bay, the site for the *Canterbury* was carefully selected for its sheltered waters avoiding any potential environmental hazards or the wreck being broken up by swells, as was the case with the *Wellington* off the southern Wellington coast.
Berkley White, courtesy Paihia Dive

17

Shipwrecks in the twenty-first century

So if by the second half of the twentieth-century radio, aircraft, and radar had removed much of the uncertainty that had earlier been such a factor in sinkings and the loss of life resulting from them, then how do things stand in the early twenty-first century? Have maritime accidents become less common? In essence, yes. Apart from the *Mikhail Lermontov*, only four large overseas ships have come ashore in New Zealand in the last 40 years — the *Pacific Charger* (Wellington 1981), *Jody F. Millennium* (Gisborne 2002), the *Tai Ping* (Bluff 2002), and the *Rena* (Tauranga 2011). All but the last were successfully refloated. Better charts, radar and most recently GPS have made our coasts much safer than in colonial days when passengers and crews could perish in their hundreds.

The fact is, the majority of maritime incidents since 2000 have been relatively minor, with smaller vessels bearing the brunt of these. Many have been the result of the same factors that imperilled so many vessels throughout New Zealand's maritime history — uncharted rocks, the unpredictability of the ocean, and the limits of human endurance.

Some have still resulted in significant loss of life, such as in the case of the commercial fishing charter *Francie*, which capsized and foundered with the loss of eight lives in November 2016 while re-entering Kaipara Harbour. As a forecast swell, coupled with the influence of an ebbing tides caused waves over the harbour bar to increase in height and break from several directions, the *Francie* was struck from behind by a large wave, causing it to capsize and sink.[1]

While in this instance the TAIC investigation found that the direct cause of the tragedy was the risk taken by the skipper in attempting to cross the bar in such conditions, the loss of the *Jubilee* with all hands the previous year ultimately remains unexplained. Only seven years old, the *Jubilee* had been fishing near Banks Peninsula for two days, when at 0030 on 18 October 2015 it ceased fishing for the night and drifted to the weather while the crew rested. Then out of the blue at 0420 the *Jubilee*'s skipper made a mayday call saying the vessel was taking on water and that the crew were taking to the life-raft. When, just before dawn, her sister-ship *Legacy* and a nearby bulk-carrier had

OPPOSITE The *Rena*, firmly grounded on Astrolabe Reef but otherwise intact. The extent of damage to the vessel and environment was yet to be realised.
New Zealand Defence Force

arrived at the mayday position, the only sign of the *Jubilee* was the empty life-raft — which had self-deployed — an oil slick, and some flotsam. A week later the bodies of the crew were discovered in the wheelhouse of the *Jubilee*, which was found upright on the seabed in about 40 metres of water. Flooding of the vessel's fishing hold was suspected as the cause of the foundering, but not categorically proven.[2]

Global positioning and modern communications technology have certainly transformed the prospects of those experiencing that sinking feeling. When the pensioner-aged fishing boat *Torea* struck an uncharted rock in Foveaux Strait on 24 August 2012, for example, the skipper's radio distress call was quickly responded to by a coordinated effort from nearby fishing vessels, the Bluff Coastguard and the Bluff Harbour pilot launch.[3] The 24 passengers and crew of three were safely rescued with little more than wet feet. The foundering of the fishing vessel *Walara-K* five years earlier resulted in a similar outcome, with the crew rescued from their lifeboat within six hours of abandoning ship, despite the incident taking place 360km off the west coast of the North Island.[4]

But technology also has its limits — as illustrated in the case of the *Pacific Charger* in chapter 14. All the technology in the world cannot cancel out human stupidity, or, indeed over-reliance or improper use of such aids. The April

The *Anatoki*, at Greymouth in August 2008, some three months after her collision with the bulk carrier *Lodestar Forest*, on the approach to Tauranga Harbour. There were no injuries and — typical of the majority of modern 'incidents' in New Zealand waters — only minor damage to both vessels that did not affect the watertight integrity of either. The damage sustained by the *Anatoki* (visible in this image) amounted to little more than a bent jack staff and bow plating, while the *Lodestar Forest* got away with slightly dented side shell plating and scratched paintwork.
Graham Ferguson

The *Hanjin Bombay* — with a decided list — re-entering Tauranga Harbour after grounding due to technical problems while negotiating the harbour channel in 2010.
Transport Accident Investigation Commission

2008 collision between the *Anatoki* and *Lodestar Forest* bulk carriers near Tauranga Harbour, while minor, was primarily put down to the masters and pilots of both vessels losing situational awareness. Equally problematic was that the *Anatoki* lacked an automatic identification system which would have made both aware of the others' position.[5]

But what of larger vessels? Here recent incidents have almost exclusively involved bulk carriers rather than cruise ships, although the latter have been involved in a few scrapes and bumps! The majority of bulk carrier incidents have also been minor — mainly groundings — though again technology and automation have frequently been a causal factor. The other core variable has been undoubtedly their sheer size — a vessel of 10,000, 20,000, or even 30,000 GRT doesn't exactly stop or turn on a dime, even with the aid of side-thrusters!

This was certainly the experience of the bulk carrier *Hanjin Bombay* on the evening of 21 June 2010 as it departed Mount Maunganui with a cargo of logs. As the bridge team took the vessel into the narrow entrance channel, increasing speed to improve the ship's steering performance, the automated engine safety control system suddenly cut in as a response to a faulty engine cooling valve. The engines stopped, just as the vessel was negotiating the turn from the channel into the harbour entrance. Unable to arrest the turn, the *Hanjin Bombay* grounded on the eastern shore of the channel, puncturing one of its ballast tanks and denting the bow hull plating.[6]

The container ship *Spirit of Resolution* would not get off quite so lightly when, three months later, it grounded on the Manukau Bar, Auckland. In this instance, the master's decision to cross the bar in light of the forecast wave height exceeding the four-metre maximum permissible was questionable. It didn't pay off. As a sudden increase in the wave height ground the vessel's progress to a halt, the current carried the *Spirit* out of the main channel, where contact with the sea-bed soon tore the rudder off, and waves began to pommel the bow. As the Transport Accident Investigation Commission effectively concluded, crossing a bar was no less dangerous in 2010 than it had been a hundred years prior.[7] The *Spirit of Resolution* wouldn't be the last bulk carrier to ground since 2000 — this 'club' also including the *Molly Manx*, which grounded in

The *Seabourn Encore* crashing into the cement carrier *Milburn Carrier II* at PrimePort Timaru in February 2017. The sheer size of modern vessels can bring a host of new problems to the fore. When the imposing profile of the *Encore* was caught by a strong southerly there was little the port facilities could do to hold the ship to the wharf. The wharf certainly came off second best — the *Encore* and *Milburn Carrier II* suffering little more than minor dents and scratches, while the Port had to seriously consider the upgrading of infrastructure.

Mac Mackintosh

Otago Harbour in August 2016; the *Leda Maersk*, in Port Chalmers in June 2018; and most recently the Panamanian-registered *Alam Seri*, which grounded at Bluff in late-November 2018.

Bulk carriers might have been more commonly involved in incidents, but cruise ships have not missed out on all the fun. The majority of these accidents have been fairly minor, such as that of the French-registered *L'Austral*, which in January 2017 pierced its hull on an uncharted rock off the Snares Islands, before striking the bottom in Milford Sound that February. Both events gave the ship the dubious honour of holding sequential incident numbers in the Transport Accident Investigation Commission's register.[8]

The *Seabourn Encore*'s experience, however, was far more exciting — at least from the perspective of land-based onlookers. While small compared to some other modern liners (Royal Caribbean's *Symphony of the Seas*, for example is 360 metres long and tips the scales at 228,000; one-and-a-half times the length and four-times the tonnage of the *Encore*) the ultra-luxurious *Seabourn Encore* still presents an intimidating 12-deck-high profile. Indeed, on 12 February 2017 this was more than its berth at Number One Wharf in Timaru could handle. The weather had been expected to turn, but as it changed earlier and more rapidly than predicted, the ship's profile began to act as something of a sail, pushing the vessel away from the wharf with such force that a number of mooring bollards were simply torn free. The resulting load on the remaining lines in turn caused them to break, sending the ship swinging across the harbour where it collided with a bulk cement carrier that was in the process of berthing at an adjacent wharf. The *Encore*'s crew had been able to engage the ship's propulsion systems in time to lessen the impact, but by that time, in addition to the wharf damage, the hull of the bulk cement carrier was holed near the waterline. Lightly damaged shell plating was the only damage for the *Encore*. The incident attests to the inadequate facilities at PrimePort Timaru. Like the majority of the smaller ports seeking to attract the lucrative cruise trade, they were playing catch-up when it came to the infrastructure necessary to support vessels of this size, and larger.[9]

'This may become New Zealand's largest shipwreck' — the *Rena*

Speaking to Radio New Zealand News on the morning of 12 October 2011, Gavin McLean offered the opinion that 'the *Rena* [was] the worst maritime environmental disaster [New Zealand has experienced].'[10] He was right. The cracks that had begun appearing in the 236-metre-long MV *Rena*, which had grounded on the Astrolabe Reef off the Bay of Plenty coast a week earlier, led to hundreds of tonnes of heavy oil being spilled off Tauranga's pristine coastal shores. McLean's notes for that radio interview — a half-page of lines written in haste — have a more literary turn: 'At the time of writing, salvage is underway in bad weather conditions. Oil and containers are in the sea, seabirds have been harmed, the crew has been taken off, the master arrested and experts report indications of hull deformation. History is in the making …' Indeed it was. The 39,000 gross ton *Rena* would be almost twice the size of New Zealand's next largest wreck, the 22,000-tonne *Mikhail Lermontov,* which sank in 1986.

The *Rena* incident effectively began just after 1000 hours on 4 October 2011 when the Liberian-registered container ship MV *Rena* departed the port of Napier, bound for Tauranga. Forwarding their estimated arrival time of 0300 the next morning to the Tauranga pilot station, within minutes of departure the vessel was at full speed. The clock was ticking.

The entry and exit to Tauranga Harbour is a complex affair at the best of times, with access to the large silty inner harbour restricted to the narrow channel between Mount Maunganui and Panepane Point. For smaller boats, navigating the tidal flow is almost a sport, with 290,000,000 tonnes of water flowing through the entrances at each tidal change generating speeds of up to 13kmph (7 knots). For large vessels, the implication is that there is a fixed window of access to and from the harbour entrance. On this particular morning, the *Rena* was set to arrive within a hair's breadth of this window closing.

When, soon after departing Napier, the *Rena*'s master learned that unfavourable currents would likely result in him missing this window, he authorised the watch-keepers to deviate from the planned course to shorten the distance and make up lost time. It was a fateful decision. As time remained elusive, the planned course to the Tauranga pilot station was adjusted, and then adjusted again, reducing clearance from two to one nautical mile. Tragically, in doing so, the second mate had failed to make an allowance for any compass error or drift. The *Rena* was tracking directly for Astrolabe Reef![11]

Soon after the master arrived back on the bridge just before 0200 hours, an intermittent echo on the radar appeared some 2.6 nautical miles dead ahead of the *Rena*. Binoculars showed nothing and according to their assumed position they were in clear water. Scanning the seas from the bridge wings similarly showed nothing. As the master turned and began to make his way to the chartroom to re-plot their position, the *Rena* — powering along at 17 knots — struck the reef. Reporting the grounding to Tauranga Harbour Control, it was established that the ship had suffered major damage with multiple compartments breached. To the astonishment of local mariners, the *Rena* was stuck fast, settling on a 10-degree list.

Ultimately the Transport Accident Investigation Commission would conclude that the grounding was entirely the result of human error, though for some it was again illustrative of the potential cost of 'flags of convenience' ships. Lower compliance costs and liabilities under international shipping conventions came at a price.[12] And right now, the price to the environment was huge. In addition to the 1700 tonnes of heavy fuel oil and 200 tonnes of marine diesel that the *Rena* was carrying, 11 of the ship's 1368 containers of cargo contained hazardous substances, including ferrosilicon, which can ignite when in contact with water.

Initial expectations of marine damage were optimistic. Maritime New Zealand's Marine Pollution Response Service mobilised immediately — as a precautionary measure — as was the transferring of the *Rena*'s fuel stores from her damaged port side to the starboard tanks.[13] But it didn't take long to realise that the damage to the vessel was more serious than realised. Within a day-and-a-half, a five-kilometre oil slick threatened wildlife and the area's rich fishing waters.[14] As expertise and equipment flooded in from around the world, however, there were signs of hope. The oil tanker *Aranuia* was soon berthed alongside to remove the *Rena*'s oil and diesel stores, which was a top priority.

Optimism soon turned to pessimism on the night of 10 October as the arrival of bad weather caused the ship to shift on the reef, sending the *Aranuia* crashing into the towering box carrier and forcing it to abandon oil transfer altogether after just 10 tonnes.[15] The resulting additional structural damage saw oil pouring out of the broken ship at five times the rate it had in the days after the grounding, ultimately causing an estimated 350 tonnes of oil being dumped into the sea.[16] Soon after, Environment Minister Nick Smith acknowledged the obvious: this was now New Zealand's worst ever maritime environmental disaster.[17]

The arrival of stormy weather had imperilled not just the *Rena*'s fuel oil stocks, but her cargo and her very structural integrity as well. Overnight, 70 containers had tumbled into the sea in the bad weather; by the following day, an additional 18. As the *New Zealand Herald*

The inevitable. During the early morning hours of 8 January 2012 the *Rena* finally broke in half as heavy seas of up to 6m battered the vessel. Officials estimated that a quarter of the 830 containers still on board were lost overboard when it broke. Things moved quickly from here. By the morning of the 10th the stern had slipped off the reef and began to submerge. By day's end only a bridge wing and part of the top hamper remained above water.
Maritime New Zealand

> **The 39,000 gross ton *Rena* would be almost twice the size of New Zealand's next largest wreck, the 22,000-tonne *Mikhail Lermontov*, which sank in 1986.**

described the situation: 'Around her, it looks like a giant has up-ended a rubbish transfer station into the blue water — now a mess of flotsam, bobbing containers and splotches of greenish oil floating around a huge, silvery oil slick trailing ominously towards the mainland.'[18] Over the following months the debris spread on water and on land. Crumpled shipping containers, tyres, milk powder, packets of noodles, and tiny plastic beads littered the coast. A dozen wheelbarrows washed up 180km away on Great Barrier Island; plywood sheets on the Coromandel coast; and at least four fridges — one as far away as Matakana.[19]

Most concerning of all, however, was the discovery of cracks on the *Rena*'s side. Exacerbated by the arrival of heavy seas, these were the inevitable result of the stern remaining afloat and shifting with the waves while the bow sat rigid on the reef. As the battering of heavy winds and rough seas took their toll, it soon became evident that the vessel was being held together only by her internal structure and the reef itself. If or rather when she actually broke apart, the environmental cost would only grow.[20] Resisting for several months, it finally happened on 8 January after a night of particularly harsh weather — the bow remaining firmly grounded on the reef, while the stern slid lower into the water, causing further debris and oil to be released as it went. Two days later, the stern was almost completely submerged, and by early April had disappeared entirely from view.[21]

By mid-2014, three-quarters of the *Rena*'s cargo had been salvaged, while by the end of the formal salvage programme a year later, some 4,593 tonnes of debris has been removed from the area.[22] The salvage operation, ending in April 2016, would ultimately cost an eye-watering $700 million — second only to the estimated $1.3 billion spent in Italy removing the cruise ship *Costa Concordia*. Clearing away the rest of the wreck would cost even more than had already been spent. After much discussion, the owners and insurers received resource consent from the government to leave a portion of the *Rena* on the reef 'in an environmentally benign state'. Not everyone agreed, however. The ruling was appealed by the Motiti Rohe Moana Trust and other parties who wanted everything removed. The Trust teamed with Forest & Bird and in mid-2018 the Environmental Court and High Court ruled in favour of the Motiti Rohe Moana Trust. Wanting to avoid further cost and to close the book on the whole sorry affair, the State made clear its intention to appeal, although at the time of writing, a decision has yet to be announced.[23]

The Motiti Rohe Moana Trust endorsed a complete ban on vessels around the Astrolabe Reef following the *Rena* incident because the absence of boating within the area immediately after the event saw marine life blossom. The trust wanted the ban to continue. The Minister of Fisheries did not. In the days, weeks, and months immediately following the *Rena*'s grounding, the environmental cost was the major issue of concern. Images of birds struggling to move with feathers covered in sticky black oil were shocking — all the more so considering the grounding occured in the midst of the breeding season for a variety of birds which fed in the waters around the ship. Five days after initially striking the reef, oil began covering the beaches along the Bay of Plenty coast, while drifting and semi-submerged containers lost over-board posed further navigational hazards, as they split open and spilt debris along the coastline. Passions ran high, and numerous public meetings quickly saw the issue become something of a political football — reportedly even resulting in a four-percent drop in support for the Key government.[24]

Although this had undoubtedly been New Zealand's worst maritime environmental disaster, ultimately fears of even greater environmental damage never eventuated. By late 2014, *Rena*-related oil wash-ups were no longer occurring. A year later, in 2015, the government-funded 'Rena Recovery' project noted in its final report that local birdlife was either stable or on the increase, and that *Rena*-related shellfish contamination was no longer at levels of concern for public or environmental health.[25] Not that the issue was put to bed, as the still on-going Motiti Rohe Moana Trust action attests. The longer-term concerns of the *Rena* grounding are now mostly confined to the continuing presence of ship and cargo debris on Astrolabe Reef and any residual chemical effects from the toxins from the cargo that was never recovered. Be that as it may, the wreck has become a beacon for many recreational divers who have embraced the opportunity that the wreck has provided with its own unique underwater experience.

By world standards, the *Rena* oil spill barely rates as a major event, but in New Zealand — a country for whom the notion of 'clean and green' has become not just a tourism slogan, but a core element of national identity — the spill garnered a uniquely different response. But it nonetheless raises an interesting question around how we can and potentially should understand the 'cost' of shipwrecks today. In the days of sail there were no fossil fuels to come ashore. The cost of wrecks was most pointedly read in terms of lives lost. Not only were the numbers staggering, but such losses

Resolve Salvage's massive RMG-500 pedestal crane barge lifting the accommodation section of the *Rena* in March 2014. Piece by piece, Resolve salvage crews removed the wreck over a six-week period using helicopters and a transport barge, with this approach chosen for the minimal impact it would have on what was already New Zealand's worst environmental disaster.
Mission Resolve

also plugged into the narrative of sacrifice that was so commonly associated with the pioneering generations. Steam took over, but those ships burned coal. When they sank off the coast, the coal sank with them (and the little that was washed ashore sometimes provided beachcombers with welcome free fuel). The loss of lives was still considerable, but to this the 'cost' of shipwrecks was often seen as part of the growing-pains of colonial development and unfortunate events that nonetheless would never deter the plucky little country from extending her commercial reach and human connection with the wider world.

All of this changed with the switch from coal to oil firing, and more particularly the more recent use of low-grade 'bunker fuel' by big ships. So although the frequency of wrecks and the loss of lives drop off significantly, there has been a profound increase in the environmental cost of wrecks. Ships have increased exponentially in size, so much so that when a big ship gets into trouble, they can spill so much more fuel and cargo than anyone in bygone days would ever have conceived possible. Modern generations commonly associate the 1989 *Exxon Valdez* oil spill in Prince William Sound, Alaska, with this shift in thinking. However, it was the loss of the oil tanker *Torrey Canyon* off the British coast in the 1960s that really brought a new awareness of the potential cost and implications of marine pollution.

The *Rena* then is uniquely poised to hold three titles. Firstly, with a registered gross weight just shy of 39,000, it has become the largest ship ever lost in New Zealand waters, coming not far off double the weight of the previous title holder, the *Mikhail Lermontov*. Prior to this, the largest ship lost in local waters was the Canadian-Australasian Line passenger liner *Niagara*, sunk off Mokohinau Island by a German-laid mine on 19 June 1940. Built in 1913, the *Niagara* was 13,415 gross.

Ecologically the *Rena* also holds the dubious distinction of being New Zealand's worst environmental maritime disaster, as too the financial title — not just for the value of the ship and cargo, but the cost of salvage as well. But in the terms of loss of human lives, fortunately she didn't win that dubious distinction. The biggest maritime death toll in New Zealand's history came with the wreck of HMS *Orpheus* on the Manukau bar on 7 February 1861, at the cost of 189 of the 250 officers and crew aboard.

There are conflicting views between those who fear the remains of the *Rena* are a ticking time-bomb, and those who argue that as an artificial reef the biodiversity around the wreck has exploded.
Ian Sherwood

Further resources

Books

Allen, Madelene Ferguson, *Wake of the Invercauld,* Exisle Publishing, Auckland, 1997

Atkinson, Neill, *Crew Culture: New Zealanders under Sail and Steam,* Te Papa Press, Wellington, 2001

Bradstock, Mike, 'Shipwrecks: the Problem of Marine Archaeology', Michael Trotter and Beverley McCulloch (eds), *Unearthing New Zealand,* GP Books, Wellington, 1989

Callan, Louise, *Shipwreck: Tales of Survival, Courage & Calamity at Sea,* Hodder Moa Beckett, Auckland, 2000

Church, Ian, *The Wreck of the Hydrabad,* Dunmore Press, Palmerston North, 1980

Clark, Charles R., *Women and Children Last: The Burning of the Emigrant Ship Cospatrick,* Otago University Press, Dunedin, 2006

Collins, Bruce E., *Rocks, Reefs and Sandbars: A History of Otago Shipwrecks,* Otago Heritage Books, Dunedin, 1995

Collins, Bruce E., *The Wreck of the Penguin,* Steele Roberts, Wellington, 2000

Doak, Wade, *The Elingamite and its Treasure,* Hodder & Stoughton, Auckland, 1969

Eunson, R. K., *The Wreck of the General Grant,* A.H. & A.W. Reed, Wellington, 1974

Fairburn, Thayer, *The Wreck of HMS Orpheus,* Whakatane Historical Society, Whakatane, 1987

Foster, Jane, *Ship on Shore: The Story of the Osprey Ship of War,* Rowfant Books, Wellington, 1982

Gibbons, Anna and Sheehan, Grant, *Leading Lights: Lighthouses of New Zealand,* Hazard Press, Christchurch, 1991

Glennie, John and Phare, Jane, *The Spirit of the Rose-Noelle,* Viking, Auckland, 1990

Gordon, Keith, *Deep Water Gold: The Story of RMS Niagara — the Quest for New Zealand's Greatest Shipwreck Treasure,* SeaROV Technologies, Whangarei, 2005

Michael Guerin, *The Mikhail Lermontov Enigma,* Chartwell Untermehmen, Blenheim, 1998

Hastings, David, *Over the Mountains of the Sea: Life on the Migrant Ships 1870–1885,* Auckland University Press, Auckland, 2006

Ingram, C.W.N., *New Zealand Shipwrecks,* 7th edn, Beckett Books Ltd, Auckland, 1990

Johnson, David, *New Zealand's Maritime Heritage,* Collins/David Bateman, Auckland, 1987

Johnson, David, *Triumph: The Ship That Hit the Lighthouse,* Dunmore Press, Palmerston North, 1981

King, Michael, *Death of the Rainbow Warrior,* Penguin, Auckland, 1986

Lambert, Max, and Hartley, Jim, *The Wahine Disaster,* A.H. & A.W. Reed, Wellington, 1969

Locker-Lampson, Steve and Francis, Ian, *Eight Minutes Past Midnight: The Wreck of the S.S. Wairarapa,* Rowfant Books, Wellington, 1981

Lambert, Max, and Hartley, Jim, *New Zealand's Shipwreck Gallery,* Rowfant Books, Wellington, 1983

Locker- Lampson, Steve, *Their Lives on Line: The Helicopter Rescue Services in New Zealand,* The Halcyon Press, Auckland, 2002

Locker-Lampson, Steve, *The Wreck Book,* Millwood Press, Wellington, 1979

Mccraw, John, *Harbour Horror,* Square One Press, Dunedin, 2001

McIntosh, Joan, *The Wreck of the Tararua,* A.H. & AW. Reed, Wellington, 1970

McLean, Gavin, *Captain's Log: New Zealand's Maritime History,* Hodder Moa Beckett, Auckland, 2001

McLean, Gavin, *New Zealand Tragedies: Shipwrecks & Maritime Disasters,* Grantham House, Wellington, 1991

Makarios, Emmanuel, *The Wahine Disaster: A Tragedy Remembered,* Grantham House/Wellington Museums Trust, Wellington, 2003

Maynard, Geoff, *Niagara's Gold,* Kangaroo Press, Kenthurst, 1996

Morton, Harry, *The Whale's Wake,* University of Otago Press, Dunedin, 1982

O'Connor, Tom, *Death of a Cruise Ship: The Mystery of the Mikhail Lermontov,* Cape Catley, Wellington, 1999

Robie, David, *Eyes of Fire: The Last Voyage of the Rainbow Warrior: Memorial Edition,* Asia Pacific Network, Auckland, 2005

Ross, John O. C., *Pride in Their Ports,* Dunmore Press, Palmerston North, 1977

Sheehan, Grant and Gibbons, Anna, *Leading Lights: Lighthouses of New Zealand,* Hazard Press, Christchurch, 1991

Simpson, Tony, *The Immigrants: The Great Migration from Britain to New Zealand 1830–1890,* Godwit, Auckland, 1997

Shears, Richard and Gridley, Isabelle, *The Rainbow Warrior Affair,* Unwin Paperbacks, Sydney, 1985

Thomson, Barry, *Deeds Not Words: The Story of the New Zealand Coastguard,* Auckland Volunteer Coast Guard Service Inc, Auckland, 1995

Warman, Mike, *The White Swan Incident: The Shipwreck That Could Have Sunk a Government,* Wairarapa Archive, Masterton, 2002

Wilkinson, Douglas, *Shipwrecks: Selected New Zealand Maritime Accidents,* Southern Press, Wellington and Dunedin, 1974

Yska, Redmer, *An Errand of Mercy: Captain Jacob Eckhoff and the Loss of the Kakanui,* Banshee Books, Wellington, 2001

Websites

Gerard Hutching, ' Shipwrecks', Te Ara — the Encyclopedia of New Zealand, updated 9-Jun-2006

URL: http://www.TeAra.govt.nz/EarthSeaAndSky/SeaAndAir Transport/Shipwrecks/en

Select maritime chronology

1642 **18 December** Abel Tasman's *Heemskerck* and *Zeehaen* anchored in Golden Bay.

1769 **6 December** HMS *Endeavour* struck a rock while leaving the Bay of Islands, but undamaged.

27 December *St Jean-Baptiste* lost three anchors off Doubtless Bay.

1772 **17 April** Marion du Fresne's *Mascarin* and *Marquis de Castries* driven out of Spirits Bay by squalls, leaving behind five anchors.

1792 Sealing industry begins with the arrival of the *Britannia* in Facile Harbour, Dusky Sound.

1795 **1 November** The supply ship *Endeavour* beached and stripped by crew in Dusky Sound — New Zealand's first marine casualty of a European-style ship.

1796 **January** Most *Endeavour* sealers/sailors leave in the *Providence*, started by the crew of the *Britannia* in 1793 — the first European ship built in New Zealand

1809 **December** *Boyd* destroyed by Maori at Whangaroa.

1831 **28 February** Brig *Industry*, her crew drunk, wrecked at Easy Harbour, Stewart Island; 17 drowned.

1834 **22 January** The *Aurora*, the first New Zealand Company immigrant ship, arrived at Britannia (modern Petone, in Wellington Harbour); vessel wrecked 27 April 1840 at the entrance to Hokianga Harbour.

29 April Barque *Harriet* wrecked off Cape Egmont and most crew killed by Maori; Captain Guard sent to obtain a ransom for the survivors, but organises revenge expedition, which in September punishes offenders and obtain release of Betty Guard and children.

May Chatham Islands Maori capture and burn the French whaler *Jean Bart* at Ocean Bay, killing 40 crew.

28 July HMS *Buffalo* blown ashore in Mercury Bay; two drowned.

9 August French ship *Comte de Paris* arrives at Pigeon Bay, Banks Peninsula.

1841 **31 March** Passengers from the first immigrant ship, the *William Bryan*, landed at New Plymouth.

3 May David Rough appointed harbourmaster of Auckland — New Zealand's first harbourmaster.

1842 **1 February** The *Fifeshire* brought the first immigrants to the New Zealand Company's Nelson settlement.

February Harbour Regulations Ordinance 1842 authorised the governor to license harbour pilots, regulate navigation and to proclaim quarantine grounds.

27 February *Fifeshire* was wrecked on Fifeshire Rock while leaving Nelson.

1846 **20 January** HMS *Driver*, the first steam ship to visit New Zealand, arrived at Auckland.

11 March HMS *Osprey* wrecked at Herekino.

1848 **23 March** The *John Wickliffe*, the first Otago settlement immigrant vessel, anchored off Port Chalmers.

1849 **29 September** Ship *Inconstant* wrecked entering Wellington; recovered but condemned and became a floating warehouse.

1850 Mercantile Marine Act 1850 placed responsibility for merchant shipping in the hands of the (UK) Board of Trade.

1 January Pencarrow lighthouse, New Zealand's first, was lit for the first time.

16 December The *Charlotte Jane* and *Randolph* arrived at Lyttelton with the first Canterbury settlers.

1851 **3 June** French corvette *L'Alcmene* wrecked between Hokianga and Kaipara; 12 drowned.

23 July Barque *Maria* lost in Cook Strait; 26 died.

24 December The *Governor Wynyard*, the first steamer built in New Zealand, launched at Auckland.

1854 Merchant Shipping Act 1854 provides the first comprehensive legislative coverage of the safety of ships, the qualifications of officers and the protection of crews

1857 **2 March** Steamer *William Denny* wrecked near North Cape.

24 December The *Governor Wynyard*, the first New Zealand-built steamship, launched at Auckland.

1860 **1 May** The *Emu* entered regular service between Auckland and North Shore, initiating steam ferry services on the Waitemata.

1861 **3 July** Steamer *Victory* wrecked Wickliffe Bay, Otago Peninsula.

1862 **29 June** Steamer *White Swan* wrecked off Castlepoint while carrying parliamentary records.

1 September Steamer *Lord Worsley* wrecked in Namu Bay, south of Cape Egmont.

November Marine Board Act 1862 placed the regulation and licensing of pilots and the construction and maintenance of lighthouses under the control of a Chief Marine Board (Steam Navigation Act 1863 also added the surveying of coastal ships to its responsibilities).

1863 **7 February** HMS *Orpheus* foundered on the Manukau bar with the loss of 189 lives — New Zealand's costliest maritime accident.

6 July Steamers *Pride of the Yarra* and the *Favourite* collided in Otago Harbour; 12 drowned.

4 September Huria Matenga rescued the crew of the *Delaware* off Pepin Island, Nelson.

1864 **3 January** Brigantine *Crafton* wrecked at the Auckland Islands, stranding five crew for 19 months.

10 May Ship *Invercauld* wrecked at the Auckland Islands; six crew drowned, three died later, 16 survivors not rescued until May 1865.

1865 21 ships totally wrecked at new gold port of Hokitika between March and December 1865.

11 May Ship *Fiery Star* caught fire and sank south of Cuvier Island; 18 rescued, 79 missing.

May Steamer *City of Dunedin* presumed to have foundered in Cook Strait after leaving Wellington on 20 May; 25 crew and 14 passengers lost.

1866 **14 May** Ship *General Grant* wrecked on Auckland Island while carrying gold and 83 passengers and crew; 10 survivors finally rescued November 1867.

11 October Chief Marine Board abolished and J.M. Balfour, Colonial Marine Engineer, assumed responsibility for lighthouses; Marine Department formed.

1867 **27 March** Immigrant ship *Montmorency* destroyed by fire at Napier.

1868 **3/4 February** Massive storm hits the South Island, wrecking the *Star of Tasmania*, *Water Nymph* and *Otago* at Oamaru, the *William Miskin* at Timaru, the *Breeze* at Le Bans Bay and the *Iona* at Lyttelton.

26 July Te Kooti and other escaped prisoners set adrift the schooner *Florence* at Waitangi, Chatham Islands and capture the schooner *Rifleman*.

1869 **14 February** Ship *St. Vincent* wrecked in Palliser Bay; 20 of 22 aboard drowned.

23/23 May Barques *Collingwood* and *Susan Jane* wrecked at Timaru; no lives lost.

1870 **23/24 July** Steamer *Tauranga* and ketch *Enterprise* collided in the Hauraki Gulf; both sank, and 18 crew and passengers from the steamer drowned.

1872 Ship *Clenmark* sailed from Lyttelton in early 1872, carrying 50 people; never seen again.

27 August Brigs *Emile* and *Scotsman* and brig *Fairy Queen* and ketch *Wanderer* wrecked at Oamaru (first two) and Timaru.

20 November The New Zealand Shipping Company was formed at Christchurch.

1874 **1 January** Immigrant ship *Surat* beached north of entrance to Catlins River after striking rocks 31 December; crew and 271 immigrants rescued by French warship *Vire* and coastal steamers the following week.

1875 **1 July** The Union Steam Ship Company of New Zealand Ltd formed at Dunedin.

1876 First overseas telegraph cable begins operating between Sydney and Wakapuaka (Nelson).

4 December Steamer *Otago* wrecked at Chaslands Mistake, south of Nugget Point, Otago.

1877 **7 August** Immigrant ship *Queen Bee* wrecked at Cape Farewell.

1878 **24 June** Ship *Hydrabad* wrecked at Waiterere Beach, between Foxton and Otaki.

22 October Immigrant ship City of Auckland wrecked at Otaki Beach.

1879 **18 February** Steamer *Taupo* wrecked while entering Tauranga Harbour; refloated but sank 29 April while under tow off Mayor Island.

1881 **29/30 April** Steamer *Tararua* wrecked off Waipapa Point; 131 lives lost.

7 November Ship *England's Glory* wrecked near the entrance to Bluff Harbour.

1882 **14 January** Ship *City of Cashmere* stranded near Timaru. Ship *Dunedin* sailed from Port Chalmers with New Zealand's first refrigerated exports.

2 May Ship *Ben Venue* and barque *Duke of Sutherland* wrecked at Timaru; the *City of Perth* was recovered and renamed *Turakina*.

1883 **3 January** The *Hurunui* was the first ship to use Lyttelton's new graving dock.

19 March The *British King*, the first full-powered steamship to trade regularly between Britain and New Zealand, arrived at Wellington.

1 September Ship *Lastingham* wrecked on Cape Jackson with the loss of 18 lives.

6 November Barque *Clyde* wrecked at Horseshoe Bay, Banks Peninsula with the loss of 18 lives.

1886 **11 April** Steamer *Taiaroa* wrecked near the entrance to the Clarence River.

12 June Ship *Lyttelton* wrecked at Timaru.

1887 **20 March** Barque *Derry Castle* wrecked on Enderby Island, Auckland Islands; 15 of 23-man crew drowned. Survivors rescued 19 July.

11 May Ship *Northumberland* wrecked on Bay View Beach, Napier. Passengers and crew rescued with great difficulty, but small steamer *Boojum* wrecked and four crew drowned while attending to the *Northumberland*.

1888 **26 January** Barque *May Queen* wrecked near Red Rock, Lyttelton Harbour.

12 June Steamer *Hawea* wrecked at New Plymouth.

23 June Steamer *Cerda* wrecked on the northern breakwater, Greymouth, severely damaging the structure.

10 July Steamer *Suva* wrecked at the entrance to Westport. HMS *Calliope* opens Calliope Dock at Auckland.

1889 **2 June** Steamer *Maitai* sank off Mercury Island; two drowned.

7 October Wreckage of barque *County of Carnarvon* found at Spirits Bay, North Cape; believed to have foundered with all hands some months earlier.

1890 **11 January** Ship *Marlborough* sailed from Lyttelton to London; went missing.

20 March Barque *Dunedin* (of frozen meat fame) sailed from Oamaru; went missing.

31 March Barque *Emilie* wrecked on Red Head, Stewart Island. Nine crew drowned or died of illness.

Barque *Assaye*, which sailed from London for Wellington on 19 February, went missing with all hands; believed to have been wrecked at The Snares.

1891 **January** Steamer *Kakanui* foundered some time after 3 January at the Macquarie Islands with the loss of 19 lives.

19 March Barque *Compadre* caught fire and went ashore at the Auckland Islands; one man died, rest of crew rescued 30 June.

1892 **9 March** Steamer *Elginshire* wrecked off Normanby Point, south of Timaru.

1893 **c. 24 February** Barque *Northern Star* believed to have foundered in a storm with all hands about 200 miles south of Awanui; hull found upside down at the South Head, Kaipara Harbour.

Late February Barque *Cowanburn* foundered in a storm with all hands off Awakino, north of Mokau.

4 September Barque *Spirit of the Dawn* wrecked on a reef off the Antipodes Island; five drowned.

29 September Barque *Evelyn* foundered near Cook Strait with the loss of all 20 hands.

1894 **18 June** Barque *Alexander Newton* wrecked on Portland Island, off Mahia Peninsula, with the loss of two lives.

29 October Steamer *Wairarapa* wrecked at Great Barrier Island; 121 drowned.

1896 **8 January** Barque *Halcione* wrecked in Fitzroy Bay, near Wellington.

1897 **16 April** Ship *Zuleika* wrecked in Palliser Bay, with the loss of 12 lives.

29 July Steamer *Tasmania* wrecked off Table Bay, Mahia Peninsula, with the loss of 13 lives.

1898 **25 August** Lifeboat *Rescue* named and launched at Sumner, Canterbury — New Zealand's first volunteer lifesaving association was formally instituted in 1904 with the formation of the Sumner Volunteer Lifeboat Brigade.

1899 **14 May** Steamer *Ohau* foundered in the vicinity of Cape Campbell, with the loss of 22 crew.

31 October Ship *Pleiades* wrecked on Akitio Beach, south of Cape Turnagain.

1900 **16 July** Steamer *Taupo* swept from its berth and wrecked at the entrance to Greymouth Harbour.

1901 **24 March** Schooner-yacht *Ariadne* deliberately wrecked off the mouth of the Waitaki River.

24 July Barque *Lizzie Bell* wrecked on Waimate Reef, Taranaki, with the loss of 12 lives.

31 October Barque *Antioco Accame* wrecked on Danger Reef, off the Shag River, Otago.

1902 **c. 24 August** Barque *Timaru* believed to have foundered in Cook Strait with the loss of all 11 hands.

28 October Steamer *Ventnor* wrecked off Omapere; 13 (some reports suggest 16) drowned.

9 November Steamer *Elingamite* wrecked on West Island in the Three Kings group; 45 died.

1903 **Late August** Ship *Loch Long* wrecked in the Chatham Islands with the loss of all hands.

7 November Barque *Northern Monarch* wrecked near Oaonui Stream, Taranaki.

11 November Barque *Ben Avon* wrecked at Cape Palliser; one death.

1904 **14 October** Barquentine *Addenda* wrecked in Palliser Bay.

15 November Barque *Kinclune* wrecked south of Kaipara Heads.

1905 **Late January** Barque *Anjou* wrecked on Cape Bristow, Auckland Island; crew survived privations and rescued mid-May.

15 June Barque *Emerald* wrecked north of the North Spit, Kaipara Heads.

20 October Barque *County of Ayr* wrecked on Danger Reef, Shag Point, Otago.

1906 **21 April** Schooner *Ronga* capsized in Pelorus Sound with the loss of six lives; wreck recovered and refurbished.

16/17 July Auxiliary schooner *Aotea* capsized in Waipiro Bay with the loss of 11 lives.

1907 **17 February** Barque *Marguerite Mirabaud* wrecked at Akatore (now Chrystals) Beach, South Otago.

7 March Barque *Dundonald* wrecked on Disappointment Island (Auckland Islands) with the loss of 11 lives; 17 survivors rescued after eight uncomfortable months.

13 June Steamer *Kia Ora* wrecked on Pirotoki Reef, Turua Point; three drowned.

14 July Barque *Woollahra* wrecked south-east of Cape Terawhiti.

Late July Barque *Constance Craig* lost at sea near Great Barrier Island with the loss of all 12 aboard.

1908 **13 March** Barque *President Felix Faure* wrecked on the North Cape, Main Island, Antipodes Islands.

18 April Four people from the fishing launch *Matanaka* drowned after the boat and the steamer *Lady Roberts* collided on Otago Harbour.

Mid-September Ship *Loch Lamond* lost with all 19 crew in the Tasman near the Northland coast.

30 October Steamer *Hawea* wrecked at Greymouth.

1909 **c. 15 January** Auxiliary brigantine *Rio Loge* believed to have foundered between Kaikoura and Banks Peninsula.

12 February Steamer *Penguin* foundered in Cook Strait with the loss of 75 lives.

27 February Ship *Forrest Hall* wrecked 25 miles south of Cape Maria Van Diemen.

1910 **4 January** Steamer *Waikare* sank in Dusky Sound.

27 June Steamer *Lauderdale* wrecked at Greymouth.

1912 **6–9 April** Dredge *Manchester* foundered in the Tasman Sea after leaving Wellington; all 25 aboard lost.

16 May Steamer *Kotuku* wrecked at Greymouth.

23 June 1912 Steamer *Star of Canada* wrecked on Kaiti Beach, Gisborne.

1913 **25 June** Steamer *Devon* wrecked at Pencarrow Head.

27 September Steamer *Tyrone* wrecked near the entrance to Otago Harbour.

1914 **7 August** Barque *Joseph Craig* wrecked inside the bar at Hokianga.

12/26 September Steamer *Kairaki* foundered off Point Elizabeth, near Greymouth, with all 17 hands.

26 September Barque *Anglo-Norman* wrecked on the North Spit, Kaipara Heads.

1916 **30 August** Steamer *Tongariro* wrecked on Bull Rock, Portland Island.

1917 **16 June** Steamer *Wairuna*, captured by SMS *Wolf* on 2 June near Sunday Island, Kermadec Group, was scuttled by the raider; the American schooner *Winslow*, which blundered on to the scene, also sunk.

3 September Steamer *Opouri* wrecked at Greymouth.

18 September Steamer *Port Kembla* sunk by mines laid by SMS *Wolf* off Cape Farewell.

1918 **December 1917/ January 1918** American schooner *Bertha Dolbeer* believed to have burned and sank with all hands off East Cape.

26 June Steamer *Wimmera* mined and sunk north of Cape Maria van Dieman; 27 killed.

24 December
Barque *Aryan* abandoned after fire east of the Chatham Islands; eight crew lost.

1921 **30 January** Auxiliary schooner *Omaka* capsized off Pencarrow Head, drowning all six crew.

8 September Schooner *Cecilia Sudden* burned and sank in the Hauraki Gulf.

13 November Steamer *Perth* wrecked at Greymouth.

1922 **31 May** Steamer *Wiltshire* wrecked on Great Barrier Island.

1924 **12 January** Steamer *Port Elliott* wrecked near Horoera Point, East Cape.

12 May Steamer *Ngahere* wrecked at Greymouth.

6 August Steamer *Ripple* foundered in Cook Strait with all 16 crew.

22 December Steamer *Konini* wrecked in Foveaux Strait.

1925 **24 May** Steamer *Cyrena* wrecked at Wanganui. Refloated but finally broken up by high seas 12 June.

1926 **10 June** Steamer *Manaia* wrecked on Slipper Island, off the Coromandel Peninsula.

2 October Steamer *Opua* wrecked at Tora, northeast of Palliser Bay.

1928 **14 July** Auxiliary schooner *Isabella De Fraine* capsized on the Hokianga bar, killing all eight crew.

1929 **16 December** Steamer *Manuka* wrecked at Long Point, south of the Nuggets, Otago; all 209 passengers and crew survived.

1931 **1 May** Steamer *Progress* wrecked in Owhiro Bay, Wellington; four crew drowned.

October Motor vessel *Kotiti* sank with five crew and two passengers some time after leaving Westport on 10 October for Foxton.

18 December Steamer *Breeze* wrecked Port Robinson; refloated but declared a constructive total loss.

1932 **27 May** Steamer *Kaponga* wrecked at Greymouth.

28 December Ten died when the launch *Doris* collided with the steamer *Tu Atu* in the inner harbour at Napier and sank.

1936 **2 February** The Lyttelton-Wellington ferry *Rangatira* was badly damaged after hitting Sinclair Head near Wellington Harbour entrance.

	18 July Steamer *Abel Tasman* wrecked at Greymouth.
1939	**19 July** Steamer *Port Bowen* stranded off Wanganui coast; defied all salvage attempts and later broken up.
	28 November Steamer *Waikouaiti* wrecked on Dog Island, Foveaux Strait.
1940	**19 June** Liner *Niagara* sunk by a German mine in the Hauraki Gulf.
	20 August Steamer *Turakina* sunk by gunfire from KMS *Orion* in the Tasman Sea; 34 lives lost.
	25 November Steamer *Holmwood* captured and sunk by KMS *Orion*, *Komet* and supply ship *Kulmerland*.
	27 November Liner *Rangitane* sunk 300 miles east of East Cape by KMS *Orion*, *Komet* and supply ship *Kulmerland*.
1941	**14 May** HMS *Puriri* mined and sunk off Bream Head; five killed.
1942	**3/4 December** Steamer *Kaiwarra* wrecked north of Motunau Island, North Canterbury.
	19 December Steamer *Wahine* ran down HMNZS *South Sea* in Wellington Harbour.
1947	**19 January** Liner *Wanganella* ran aground on Barrett Reef, Wellington; refloated 18 days later.
1950	**28 December** Passenger launch *Ranui* capsized at North Rock, Mount Maunganui; 22 drowned.
1957	**15 May** Auxiliary scow *Lena* foundered near Waiheke Island; six drowned.
1959	**24 November** Coaster *Holmglen* foundered north of Waitaki River; 15 drowned.
1963	**20 September** Coaster *Holmbank* wrecked near Peraki Bay, Banks Peninsula.
1966	**23 May** Collier *Kaitawa* lost near Cape Reinga with all 29 hands.
1968	**10 April** Ferry *Wahine* foundered in Wellington Harbour; 51 lives lost.
	13 June Coaster *Maranui* foundered 25 miles east of Great Mercury Island; nine drowned.
1971	**19 June** The *Columbus New Zealand*, the first all-container ship to visit New Zealand, berthed at Wellington.
1975	**8 February** Freighter *Union East* abandoned 300 miles off East Cape with the loss of one life; the ship sank 12 February.
	3 September Sixteen died after abandoning the burning freighter *Capitaine Bougainville* off the Northland coast; hulk scrapped at Auckland in 1977.
1976	**31 July** Royal New Zealand Coastguard Federation formed.
	15 September The *Rangatira* completed the last Lyttelton–Wellington passenger ferry service.
1978	**11 June** Freighter *Kemphaan* caught fire NE of Le Bons Bay, Banks Peninsula; two crew killed. Wreck later sunk as a target by RNZAF jets.
1981	**21 May** Freighter *Pacific Charger* grounded near Baring Head, near the entrance to Wellington Harbour; refloated 5 June.
1984	**29 June** Steam-trawler *Hawea* sank WNW of Kahurangi Point.
1985	**10 July** French secret agents bombed the *Rainbow Warrior* at Auckland; one crewman killed.
1986	**16 February** Liner *Mikhail Lermontov* sank in Gore Bay, after striking rocks off Cape Jackson, Marlborough Sounds; one crewman drowned. Largest ship ever lost in New Zealand waters.
	2 July Police launch *Lady Elizabeth II* foundered off Barrett Reef, Wellington Harbour, with the loss of two lives.
1989	Ports reform saw harbour boards abolished and their commercial responsibilities transferred to publicly owned port companies.
	February Fishing vessel *Sankichi Maru No. 18* caught fire in the Tasman Sea about 300 miles west of the Manukau Harbour; 15 fishermen lost and the drifting hulk sank on 13 February.
	30 September Upturned trimaran *Rose-Noelle* broke up off Great Barrier Island after drifting upside-down in the Pacific for 119 days. Amazingly, all four crew members survived.
1994	**6 July** Stern trawler *Amaltal Challenger* sank in Cook Strait.
1996	**29 December** Container ship *Sydney Express* and fishing vessel *Maria Luisa* collided near the entrance to Wellington Harbour; five fishermen drowned.
1998	**6 October** Korean fishing vessel *Dong Won 529* went ashore near the eastern tip of Breaksea Island, near Stewart Island; sank two days later.
1999	**12 February** Woodchip carrier *Prince of Tokyo* grounded off the North Mole, Otago Harbour, while leaving port; refloated 14 February.
2000	**17 March** Stern-trawler *Seafresh 1* sank in Hanson Bay, Chatham Islands, after catching fire NE of the islands on 9 March.
2002	**6 February** Bulk carrier *Jody F. Millennium* grounded while leaving Gisborne Harbour; refloated 18 days later.
	8 October Bulk carrier *Tai Ping* grounded in fog at Bluff, sustaining serious damage.
	14 July Roll-on, roll-off freighter *Kent* holed and seriously damaged while berthing in Wellington Harbour.
2004	**18 December** Ferry *Tiger III* grounded and became a total loss beneath the Cape Brett lighthouse. Three crew and 59 passengers were rescued.
2005	**4 January** Passenger ferry *Quickcat* and charter boat *Doctor Hook* collided in Motuihe Channel, injuring several aboard the *Doctor Hook*. One passenger later died of injuries and the *Doctor Hook* was written off.
2006	**13 May** Trawler *Kotuku* capsized in Foveaux Strait, drowning six.
2007	**7 March** The fishing vessel *Walara-K* flooded and sank 195 nautical miles off the west coast of the North Island. Three crew members abandoned ship and were rescued.
2008	**27 July** Fishing vessel *San Cuvier* dragged its anchor and grounded on rocks near Tarakeha Point, Bay of Plenty, with the loss of two crew members. The remaining crew members survived by scaling the rocks close to the stricken vessel.
2010	**18 September** Coastal container ship *Spirit of Resolution* grounded on Manukau Bar, shearing off most of its rudder. With the aid of its bow thrusters the ship was manoeuvred into deeper water, then towed to Lyttelton for repair.
2011	**5 October** MV *Rena* grounded on Astrolabe Reef, resulting in the total loss of the vessel.
2012	**15 March** Fishing vessel *Easy Rider* capsized and foundered in Foveaux Strait after being struck by a large wave. Of the three crew and six passengers on board at the time, the only survivor was one crew member.
2015	**18 October** Loss of the fishing vessel *Jubilee* and all hands, 12 nautical miles off the Rakaia River mouth.
2016	**18 January** Restricted-limits passenger vessel *PeeJay V*, on an excursion from Whakatane to White Island with 60 passengers and crew, caught fire and sank. There were no serious injuries.
	26 November Charter fishing vessel *Francie*, foundered and capsized on Kaipara Harbour bar, with the loss of eight lives.

Notes

1 Taming treacherous coasts

1. Max Lambert and Jim Hartley, *The Wahine Disaster,* A.H. & A.W. Reed, Wellington, 1969, p. 37.
2. Gerard Hutching. 'Shipwrecks', Te Ara — the Encyclopedia of New Zealand, updated 9-June 2006, URL: http://www.TeAra.govt.nz/EarthSeaAndSky/SeaAndAirTransport/Shipwrecks/en.
3. Wade Doak, *The Burning of the Boyd,* Hodder & Stoughton, Auckland, 1984.
4. Cited in Augustus Earle, *A Narrative of Nine Months' Residence in New Zealand in 1827: Together with a Journal of Residence in Tristan D'Acunha,* Longman, 1832, p.45.
5. Hugh Murray, *Adventures of British Seamen in the Southern Ocean: Displaying the Striking Contrasts which the Human Character Exhibits in an Uncivilized State,* Constable and Company, 1827, p. 348
6. Gavin McLean, *Otago Harbour: Currents of Controversy,* Otago Harbour Board, Dunedin, 1985, p. 306.
7. *North Otago Times,* 27 Nov 1866.
8. Marine Department Report for 1869–70 and 1870–1, *AJHR,* 1867, Vol. II, E-6, p. 3.
9. McLean, *Otago Harbour,* p. 35.
10. Neill Atkinson, *Crew Culture: New Zealand Seafarers Under Sail and Steam,* Te Papa Press, Wellington, 2001, p. 51.
11. http://www.taic.org.nz/index.html
12. Gavin Mclean, *Captain's Log,* Hodder Moa Beckett, Auckland, 2001, p. 206.
13. Mike Pryce, 'Nautical News', *NZMN* 53/4, 2005, p. 201.
14. Ibid, 49/1, 2000, pp. 30–1.
15. MSA media releases, 1 and 3 Apr 2003.
16. *New Zealand Herald (NZH),* 26 Jun 2003.

2 Drunk, deranged or dangerous

1. Marine Department Report for 1869–70 and 1870–1, *AJHR,* 1871 Vol. II, G-6, p. 5.
2. Henry Brett, *White Wings: Fifty Years of Sail in the New Zealand Trade 1850 to 1900,* Vol. I, The Brett Publishing Co., Auckland, 1924, pp. 238–9.
3. For an exhaustive account, see Thayer Fairburn, *The Orpheus Disaster,* Whakatane & District Historical Society, Whakatane, 1987.
4. Elsie Locke, 'From Forecastle to Gaol', unpublished MS in the possession of the New Zealand Maritime Union.
5. Atkinson, *Crew Culture,* p. 103.
6. Locke, 'From Forecastle to Gaol', p. 11.
7. Locke, pp. 97–9, quoted in Atkinson, *Crew Culture,* p. 104.
8. *Otago Witness,* 13 Jul 1851.
9. See Bruce E. Collins, *The Wreck of the Surat,* Otago Heritage Books, Dunedin, 1991.
10. *Bruce Herald,* 6 Jan 1874.
11. Collins, *Surat,* p. 23.
12. An azimuth is an angular measurement in a spherical coordinate system, such as in the tracking of stars for the purpose of navitgation.
13. Union Steam Ship Company (USSCo) minutes, 18 Aug 1883, HC (Hocken Collections).
14. USSCo minutes, 18 Aug 1883, HC.
15. Robert Strang to Angus Cameron, 3 Mar 1892, Cameron Papers, MS 1046/4, HC.
16. C. Monson to James Mills, 25 Aug, 1900, James Mills Papers, HC.
17. Richardson & Co. minutes, 19 Dec 1908, WCC Archives.

3 Bar harbours — river ports

1. Philip Ross May, *Hokitika: Goldfields Capital,* Pegasus Press, Christchurch, 1964, p. 10.
2. Ibid, p. 11.
3. *West Coast Times,* 10 May 1867.
4. Undated clipping, James Mills Papers, USSCo archives, HC.
5. *Dominion,* 30 May 1932.
6. Ibid, 8 Jun 1932.
7. *Evening Post,* 1 Oct 1898.
8. Ibid, 12 Oct 1898.
9. Ibid, 16 Mar 1899.

4 Exposed coasts — roadstead ports

1. *Oamaru Times,* 2 Apr 1867.
2. Ibid, p. 50.
3. *Beginnings: Early History of North Otago, Oamaru Mail,* Oamaru, 1934 and 1978, p. 132.
4. *North Otago Times,* 29 Aug 1873.
5. *Oamaru Times and Waitaki Reporter,* 19 Mar 1867.
6. Bruce E. Collins, *Rocks, Reefs and Sandbars: A History of Otago Shipwrecks,* Otago Heritage Books, Dunedin, 1995, p. 45.
7. Collins, ibid, p. 50.
8. R.A. Chapman, 'Canterbury's Nineteenth Century Specially-designed Lifeboats', *NZMN* 51 / 1, 2002, p. 12.

5 Hot times in cold climes

1. *White Wings,* p. 48.
2. *North Otago Times,* 19 Mar 1890.
3. Oamaru Harbour Board, *Annual Report* 1890, NOM.
4. Not everyone believed that. According to a paper in the *Transactions and Proceedings of the New Zealand Institute,* 1898, 'On only one occasion (in April 1892) is the fact mentioned that observations had been made on the temperatures of the air and the sea in the vicinity of icebergs. Mr. Barthorp, however, informs me that such observations have been regularly made, and with the disappointing result that no reliable information evidencing the proximity of an iceberg is to be obtained by thermos-metrical observations, but that in the dark, and in foggy and thick weather, the only means available to safeguard a ship against dangerous collisions with icebergs is to keep always a most vigilant look-out. http:// rsnz.natlib.govt.nz/volume/rsnz_31/rsnz_31_00_008520.html
5. *White Wings,* Vol. 1, p. 56.
6. A.H. & A.W. Reed (eds), *Castaway on the Aucklands: The Wreck of the Grafton from the Private journals of Thomas Musgrave, Master Mariner,* A.H. & A.W. Reed, Wellington, 1943, p. 33.
7. Francois Rayna I, *Wrecked on a Reef, or Twenty Months Among the Auckland Isles,* facsimile edition, Steele Roberts, Wellington, 2003, pp. 22–3.

8. Madelene Ferguson Allen, *Wake of the Invercauld: Shipwrecked in the sub-Antarctic: A Great-Grand-Daughter's Pilgrimage,* Exisle Publishing, Auckland, 1997, p. 152.
9. http://www.teara.govt.nz/EarthSeaAndSky/ SeaAndAirTransport/Castaways/1/en
10. Henry Armstrong, 'Cruise of the brig Amherst', *New Zealand Government Gazette,* Province of Southland 6, 9 (11 Apr 1868), p. 52.
11. Rowley Taylor, *Straight Through From London: The Antipodes and Bounty Islands, New Zealand,* Heritage Expeditions New Zealand Ltd, Christchurch, 2006, p. 171.
12. E.R. Martin, *Marine Department Centennial History 1866–1966,* Government Printer, Wellington, 1969, p. 87.
13. http://www.doc.govt.nz/templates/page.aspx?id+34229

6 'Terrible engines of destruction'

1. *ODT,* 10 Jul 1863.
2. The collision has been described in John McCraw, *Harbour Horror: Story of the Tragic Collision of Two Ferries on Otago Harbour in 1863,* Square One Press, Dunedin, 2001.
3. McCraw, p. 15.
4. *ODT,* 10 Jul 1863.
5. McCraw, pp. 71–6.
6. Colin Thompson, 'The *Taranaki-Waipiata* Collision, *NZMN* 37/4, 1987, pp. 154–6.
7. *Southern Cross,* 20 Jun 1950.
8. Nautical adviser to the secretary of marine, 7 Jul 1950, M1/13.2640, Archives New Zealand (ANZ).
9. Ibid.
10. Ibid, 10 Jul 1950, MI, 13.2640, ANZ.

7 'A thrill of horror throughout the Empire'

1. David Hastings, *Over the Mountains of the Sea: Life on the Migrant Ships 1870–1885,* Auckland University Press, Auckland, 2006, p. 70.
2. Ibid.
3. Sir Henry Brett, *White Wings,* Vol. 1, the Brett Publishing Co, Auckland, 1924, p. 62.
4. For a detailed account of the *Cospatrick,* see Charles R. Clark, *Women and Children Last: The Burning of the Emigrant Ship Cospatrick,* Otago University Press, Dunedin, 2006.
5. Clark, p. 55.
6. Brett, p. 63.
7. Ibid.
8. Clark, p. 88.
9. *Hawke's Bay Herald,* 30 Mar 1867.
10. *Evening Post,* 2 Aug 1918.
11. *Press,* 12 Dec 1955.
12. Secretary of marine to superintendent mercantile marine, 7 Jul 1955, AAPR Ace W3282, 13/2712, Pt 1, *Pateke,* ANZ.
13. Report of Court of Inquiry into the Fire aboard the *Holmburn,* AAPR, Ace W3350, 11, ANZ.
14. Ibid.
15. Ibid. Crews of small vessels had been reluctant to keep such a watch because of the demand it placed on stretched resources, especially when the port was a homeport for several crew who naturally preferred to spend the night at home.
16. Undated press clipping, ABPL, 7457, Ace W4932, 77, 39/4/271, part 1, ANZ.
17. Captain J.R. Thomas, statement, TR 1, Ace W25552, 39/4/271, part 1, ANZ.
18. Ibid.
19. Ingram, *New Zealand Shipwrecks,* p. 451.
20. P.M. Kershaw, report, TR 1, Ace W25552, 39/4/271, part 1, ANZ.
21. 'Whilst he took measure to cut off air supply to contain the fire, but closed no fuel tank valves, the lack of effective action to fight the fire brings into question the Chief Engineer's understanding of the smothering gas system and his early report to the Master that the fore was out of control suggests that he reacted somewhat impulsively to the emergency,' the secretary of transport advised the minister on 13 November 1975. 'Looking back on the events it must be said that the decision [to abandon ship] was made prematurely without full knowledge of the circumstances.' TR 1, Ace W25552, 39/4/271, part 1, ANZ.

8 Two of the worst

1. Information for this chapter is based on newspaper clippings and original documents from Union Company papers held by the Wellington City Archives. Also useful was Steve Locker-Lampson and Ian Francis, *Eight Minutes Past Midnight: The Wreck of the s.s. Wairarapa,* Rowfant Books, Wellington, 1981.
2. The 1895 Marine Department report put the toll at 126; *AJHR* 1895, H-29, p. 4.
3. *Otago Daily Times,* 3 Nov 1894.
4. Ibid, 14 Nov.
5. Ibid, 3 Nov.
6. Press clipping, 'Wairarapa File', USSCo papers, op. cit.
7. *Otago Daily Times,* 3 Nov 1894.
8. *Auckland Star,* 29 Oct 1894.
9. 'Report of the Court of Inquiry', printed as a supplement to the *Otago Witness,* 13 Dec 1894.
10. Ibid.
11. Ibid.
12. Ibid.
13. James Mills, 'Finding of the Court of Inquiry in the Matter of the Wreck of S.S. *Wairarapa* Together With Memorandum Prepared by Managing Director for the Consideration of the Board and List of Lost and Saved', 10 Jan 1894, USSCo papers, op. cit.
14. Ibid.
15. Ibid.
16. Steve Locker-Lampson and Ian Francis, *The Wreck Book,* Rowfant Books, Wellington, 1979, pp. 96–7.
17. Angie Belcher, 'Great Barrier Island', *Dive New Zealand* Jun/Jul 1998, http://www.divenewzealand.com/articles.asp?sid=2 83
18. *Evening Post,* 13 Feb 1909. Most direct quotes in this chapter are drawn from the *Post* on that and subsequent days.
19. The widely quoted figure of 75 is wrong. For this and other matters, see Bruce E. Collins, *The Wreck of the Penguin,* Steele Roberts, Wellington, 2000.
20. http://wellington.govt.nz/services/cemeteries/pdfs/penguinwalk.pdf
21. For the *Rio Loge* theory. See Allan A. Kirk, *Express Steamers of Cook Strait,* A.H. & A.W. Reed, Wellington, 1968, pp. 140–1 and Collins, pp. 87–94.

9 Fraud, vandalism and terriorism

1. The term 'coffin ship' is particularly associated with the Irish immigrant trade in the mid-nineteenth century.
2. Ingram, p. 62.
3. Robert Fulton Valpy, *Medical Practice in Otago and Southland in the Early Days,* Otago Daily Times and Witness Newspaper Co., Dunedin, 1921, p. 67.
4. Quoted in 1.J. Farquhar, 'The Harbour Steam Company and the Origins of the Union Company', *NZMN,* 26/ 4, 1975. p.114
5. Margaret de Jardine, *Shipwrecks on and off the Taranaki Coast,* author, New Plymouth, n.d., p. 107.
6. A.B. Scanlan, *Harbour at the Sugar Loaves,* Taranaki Harbour Board, New Plymouth, 1975, p. 61.
7. de Jardine, p. 109.
8. M. Newman to James Mills, 20 June 1888, Mills Papers, USSCo archives, WCA.
9. The wreck of the *Ariadne* is described in the *Oamaru Mail* of 25 March 1901. The best modern summary of the case is Louise Callan, *Shipwreck: Tales of Survival, Courage & Calamity at Sea,* Hodder Moa Beckett, Auckland, 2004, pp. 68–79.
10. *ODT,* 14 Apr 1901.
11. Ken Catran, *Hanlon: A Casebook,* BCNZ Enterprises, Auckland, 1985, p. 73.
12. ABFK, 7194, Ace W4948, 172, 42/2/7, ANZ.
13. Ibid. The last pieces of the *Hautapu* were lifted on 29 March. Final costs were $24, 900 to the military and $10, 400 to the WHB.
14. David Robie, *Eyes of Fire: The Last Voyage of the Rainbow Warrior, Memorial Edition,* Asia Pacific Network, Auckland, 2005, p. 116.
15. David Lange, *Nuclear Free — the New Zealand Way,* Penguin, Auckland, 1990, p. 122.
16. G.C. Wright, 'Salvage of the 'Rainbow Warrior'', NZMN, 53 / 1, 2005, p. 15.
17. Ibid., p. 21.
18. Robie, *Eyes of Fire,* p. 164.
19. Dave Abbott, 'Northern Wrecks Tour — Six of the Best', *Dive New* Zealand, Jun/Jul 2004: http://www.divenewzealand.com/articles.asp?sid=l21
20. Wright, 'Salvage of the 'Rainbow Warrior'', p. 22.

10 Casualties of war

1. *Evening Post,* 1 Jul 1918.
2. Ibid, 28 Jul 1918.
3. For the story of the German raiders, see Karl August Muggenthaler, *German Raiders of World War II,* Robert Hale, London, 1978 and Paul Schmalenbach, *German Raiders: A History of Auxiliary Cruisers of the German Navy 1895–1945,* Patrick Stephens, Cambridge, 1979.
4. For the *Niagara story,* see Keith, Gordon, *Deep Water Gold: The Story of RMS Niagara — the Quest for New Zealand's Greatest Shipwreck Treasure,* SeaROV Technologies, Whangarei, 2005 and Jeff Maynard., *Niagara's Gold,* Kangaroo Press, Kenthurst, 1996.
5. Press clipping, M1, 13.2450, ANZ.
6. Reginald Kerr, deposition, 21 Jun 1940, Ml, 13 .2450, ANZ.
7. William Hart, deposition, 21 Jun 1940, ibid.
8. Ibid.
9. *NZH,* 22 Jul 1940.
10. Tim Cashman, ' Niagara 2000', in *Dive New Zealand,* Apr/May 2000, http ://www.divenewzealand .com/artic I es.asp? sid=54.
11. Press clipping, Ml, 13 / 2455, ANZ.
12. *Dominion,* 9 Jan 1941.
13. C.L. Spencer to secretary of marine, 19 Dec 1940, Ml, 13/2460, ANZ.
14. Chief of naval staff to the minister of marine, 21 Jan 1941, N 1, 1 6/8 .28, ANZ. In his deposition of 16 January 1941, Captain Miller said' I realised it would be useless to attempt to use the wireless as with my plant I had never been able to contact Wellington from this distance and Chatham Island Station did not open until 9 a.m.'.
15. Estimates of the number of people aboard the *Rangitane* vary slightly. See Muggenthaler, p. 81.
16. F.F. Howells, 10 Mar 1946, IA 1, 181 / 19/4, War History — Narratives Merchant Shipping — Seamen' s Accounts, ANZ.
17. L.N. Sowerby, ibid.

11 Conference Lines casualties

1. *Evening Post,* 26 Aug 1913.
2. Ibid.
3. Ibid.
4. *Dominion,* 27 Aug 1913.
5. *Evening Post,* 27 Aug.
6. Ibid, 2 Sep 1913.
7. Solicitor-general to attorney-general, 5 Sep 1913, Ml, 13/ 13, *Devon Inquiry,* ANZ.
8. *Dominion,* 19 Sep 1913, clipping Ml, 13/ 13.
9. *Dominion,* 27 Aug 1913.
10. Ibid, 28 Aug 1913.
11. Bertram Hayward, deposition, 21 Jun 1922, Ml, 13/ 821, pt 2, ANZ.
12. *Evening Post,* 2 Jun 1922.
13. Ibid.
14. Ibid.
15. *Dominion,* 5 Jun 1922, Ml, 13/ 821, pt 2, ANZ.
16. *Gisborne Times,* 14 Dec 1924.
17. See Ian Farquhar, '"Port Bowen" — the Ship that Came and Never Left', *NZMN,* 54/ 2, 2006, pp. 58–68.

12 Floating Jonahs and hoodoo ships

1. Gavin Dobie, 'Sailing Scow "Echo" 100 Years Old', *NZMN* 53/ 2, 2005, p. 110.
2. The Blenheim collector of customs sometimes reported minor accidents along with complaints about non-compliance with the paperwork. ' Neither the Master nor the owners took any steps to advise me of the mishap,' he complained in April 1939 after the ship spent a week on the Wairau bar, 'as apparently they consider that as the vessel's seaworthiness was not in any way affected, they are not required to report the stranding.' AAPR, Ace W3282,8, 13/2305 Part 1, ANZ.
3. *Evening Post,* 17 Oct 1960.
4. *Dominion,* 28 Nov 1932.
5. Ibid.
6. Crown solicitor to the secretary of marine, 13 Dec 1932, Ml, 13/ 1934, ANZ.

7. 'The report received from the Court is disappointing,' Harold Ruegg told the secretary on 10 October 1956. 'Although the report seems to accept the Master's evidence that the ship actually got through the passage without difficulty, the evidence of the Second Mate and Mate was to the effect that she did not get through the passage and experienced all the difficulty in the passage.' AAPR, W3282, 9, 13/2738, Part 1, ANZ.
8. TAIC, 1.3.4.
9. TAIC, 2.33. and 2.3.4.
10. TAIC, 2.2.2. 'It is difficult to explain why neither the master nor the third mate on the *Sydney Express* saw the white masthead light on the *Maria Luisa*. The evidence of the Beacon Hill operator, the crew of the *Soundsgood* and the deck-hand on the *Maria Luisa* suggests that the light was burning brightly. The masthead light on the *Maria Luisa* stands well above the sidelights and the glow of the deck lights. It is likely that this, coupled with their preconception that the *Maria Luisa* was a sailing vessel, caused them to miss the masthead light.'
11. TAIC, 1.1.35.
12. TAIC, 1.1.35.
13. TAIC, 1.1.37.
14. TAIC, 1.137.
15. MSA, p. 6.
16. MSA, p. 4.
17. TAIC, 3.13.
18. TAIC, 3.18.
19. TAIC, 3.21.
20. TAIC, 3.29.
21. Webb, 'Review of the Submission to the Maritime Safety Authority', p.1.

13 Coasters in crisis

1. Information is drawn from the Canterbury Steam Shipping Company files held by the Wellington City Archives.
2. Press clipping on TRl, Ace W2552, 13/ 274/1 a, ANZ.
3. Helen Beaglehole, *Lighting the Coast: A History of New Zealand's Coastal Lighthouse System,* Canterbury University Press, Christchurch, 2006, p. 254.
4. Henry Williams, deposition, 24 Nov 1959, AAPR, Ace W3282, 10, 13/ 28669, Part 2, ANZ.
5. Rod Donald, statement, 3 Dec 1959, AAPR, Ace 3350, 10, ANZ.
6. *Press,* 28 Nov 1959.
7. J. lnkster, statement, 27 Nov 1959, AAPR, Ace 3350, 10, ANZ.
8. Henry Williams, statement, AAPR, Ace 3350, 10, ANZ.
9. J.C. Winders to secretary of marine, AAPR, Ace W3282, 10, 13/ 2860, part 2, ANZ.
10. Allan McKay, deposition, 2 Dec 1959, AAPR, Ace 3350, 10, ANZ.
11. Eric Tutty, deposition, 2 Dec 1959, AAPR, Ace 3350, 10, ANZ.
12. Harry Burnett, deposition, 1 Dec 1959, deposition, 2 Dec 1959, AAPR, Ace 3350, 10, ANZ.
13. Rowland Masterman, deposition, M 1, 12/ 1113.
14. J.D. Carrick, evidence, AAPR, Ace 3350, 10, ANZ.
15. Patrick Gordon, evidence, AAPR, Ace 3350, 10, ANZ.
16. Alex Grieve, evidence, AAPR, Ace 3350, 10, ANZ.
17. Cosmo Keith, evidence, AAPR, Ace 3350, 10, ANZ.
18. Louise Callan, *Shipwreck: Tales of Survival, Courage & Calamity at Sea,* Hodder Moa Beckett, Auckland, 2000, p. 140.
19. E.F. Rainbow to minister of marine, 30 Mar 1960, Ml, 12/ 113, ANZ.
20. Thomas Edge, deposition, AAPR, Ace W3282, 10, 13/2929, part 1, ANZ.
21. E. Milroy to minister of marine, 29 Jul 1966, AAPR, Ace W3282, 10, 13/2929, part 1, ANZ.
22. Thomas Edge, deposition, AAPR, Ace W3282, 10, 13/ 2929, part 1, ANZ.
23. Milroy to minister, ibid.
24. N.L. Merrick, statement, 15 Jun 1966, AAPR, Ace W3282, 10, 13/2929/2, Part 1.
25. Milroy to minister, ibid.
26. Roger Wincer's website http://users.iconz.co.nz/rwincer/kaitawa.htm disagrees with Milroy's analysis.
27. *Dominion,* 24 Nov 1966.
28. *NZH,* 12 Dec 1995.
29. Ibid, 11 Aug 1954.
30. Ibid, 6 May 1953.
31. Ibid, 29 Nov 1967.
32. Ibid, 8 Oct 1968.
33. Ibid, 12 Dec 1995.
34. Ibid, 18 Jun 1968.
35. *Auckland Star,* 21 Jan 1969
36. W.J. Scott to Capt. T. Wahlstedt, 17 Jun 1968, ABPL, 7458, Ace 4932, 163, ANZ.

14 Close calls

1. Peter Plowman, *Passenger Ships of Australia and New Zealand, Volume II 1913–1980,* William Collins, Auckland, 1981, p. 112.
2. Max Hodgson, ' Fifty Years Ago: The Stranding of the *Wanganella',* NZMN 46/ 2, 1997, p. 66.
3. Gavin McLean, *Moeraki: 150 Years of Net and Plough Share,* Otago Heritage Books, Dunedin, 1986, p. 74.
4. TAIC Report, 02-201, *Bulk log carrier Jody F. Millennium grounding at Gisborne 6 February 2002,* 12 May 2003, p. 2.
5. TAIC Report 02-201, p. 1.
6. Mike Pryce, ' Nautical News, *NZMN* 51/2, 2003, p. 79.
7. TAIC Report 02-201, p. 17.
8. Quoted in Pryce, *NZMN* 51/3, 2003, pp. 148–9.
9. MSA press release, 8 Apr 2003.
10. Pryce, p. 148.
11. TAIC Report 02-201, p. 17.
12. TAIC report 02-206, *Bulk Carrier Tai Ping, Grounding, Bluff Harbour,* 12 May 2003, p. 20.
13. Pryce, *NZMN* 52/2, p. 102.

15 'Rocks ahead! ... Rocks astern!

1. For accounts of the *Wahine* disaster, see Max Lambert and Jim Hartley, *The Wahine Disaster,* A.H. & A.W. Reed, Wellington, 1969 and Emmanuel Makrios, *The Wahine Disaster —A Tragedy Remembered,* Grantham House Publishing in association with the Wellington Museums Trust, Wellington, 2003.
2. Allan A. Kirk, *Express Steamers of Cook Strait,* A.H. & A.W. Reed, Wellington, 1968, p. 126.
3. Susan Butterworth, *More than Law and Order: Policing a Changing Society, 1945–1992,* University of Otago Press, Dunedin, 2005, pp. 122–5.

4. For accounts of the *Mikhail Lermontov* disaster, see Michael Guerin, *The Mikhail Lermontov Enigma,* Chartwell Untermehmen, Blenheim, 1998, and Tom O'Connor, *Death of a Cruise Ship — The Mystery of the Mikhail Lermontov,* Cape Catley, Whatamango Bay, 1999.
5. *NZH,* 21 Feb 1996.
6. Gavin McLean, *Rocking the Boat? A History of Scales Corporation Limited,* Scales Corporation, Christchurch, 2001 p. 291.
7. http ://www.nzherald.co.nz/section/story.cfm?c_id=1&Object!D=10368542
8. http://www.nzmaritime.co.nz/lermontov.htm
9. Pete Mesley, 'Lermontov 99', *Dive New Zealand* Apr/May 1999, b.lli2JLwww.divenewzealand.com/articles.asp?sid=367
10. Dave Moran, 'Mikhail Lermontov Celebrates Her 20th Birthday', *Dive New Zealand* Apr/May 2006, http://divenewzealand.com/articlers.asp?sid=722
11. Mike Pryce, 'Nautical News', *NZMN* 54/1, 2005, p. 24.
12. Ibid., p. 25.
13. Pryce, *NZMN* 54/3, 2006, pp. 136–7.

16 From horror to heritage

1. See, for example, Eric Ramsden, *Busby of Waitangi,* Wellington and Dunedin. 1942, p. 72, and A.D. Mcintosh (ed), *Marlborough: A Provincial History,* Blenheim. 1940, pp. 21, 24, 44.
2. Kynan Gentry, *History, Heritage, and Colonialism: Historical consciousness, Britishness, and cultural identity in New Zealand, 1870–1940,* Manchester University Press, 2015, pp. 67–73.
3. Russell Duncan, 'Following the Tracks of Captain Cook', *Transactions and Proceedings of the New Zealand Institute,* 35 (1902), pp. 43–4.
4. http://www.oceans.com.au/oektsia.html
5. 'Gang has the Tarlton haul, thief claims', http://www.stuff.co.nz/national/2838670/Gang-has-the-Tarlton-haul-thief-claims
6. E.V. Sale, *Kelly: The Adventurous Life of Kelly Tarlton,* Heinemann Reed, Auckland, 1988. p. 55. See also Steve Locker-Lampson, *Throw Me the Wreck Johnny: Memories of Kelly Tarlton,* Halcyon, Auckland, 1996.
7. Mike Bradstock, 'Shipwrecks: The Problems of Marine Archaeology', in Michael Trotter and Beverley McCulloch (eds), *Unearthing New Zealand,* GP Books, Wellington, 1989, p. 106.
8. *Historic Places in New Zealand,* Sep 1983, p. 31.
9. Andrew Dodd, 'Opportunities for Underwater Archaeology in New Zealand', *Archaeology in New Zealand,* 46:3 (2003), pp. 151-60.
10. *Dominion,* 18 Jul 1998.
11. David Churchill, 'The maritime Archaeological Association of New Zealand MAANZ (Inc)', http://www.maanz.wellington.net.nz/projects/amipap.html
12. Heritage New Zealand, 'Shipwrecks and Underwater Archaeological Sites', Heritage New Zealand, July 2016.
13. 'Seabed survey reveals 108-year-old shipwreck', https://www.nzherald.co.nz/nz/news/article.cfm?c_id=1&objectid=12115648; 'Mystery around wreck find off Wellington coast', https://www.stuff.co.nz/dominion-post/news/105880441/mystery-around-wreck-find-off-wellington-coast
14. 'The Daring - what history lies in a 153-year-old shipwrecked schooner?', https://www.stuff.co.nz/auckland/109415169/the-daring--what-history-lies-in-a-153yearold-shipwrecked-schooner

17 Shipwrecks in the twenty-first century

1. TAIC, MO-2016-206.
2. 'Three fishermen who died on ill-fated FV Jubilee were trapped in wheelhouse, inquiry finds', https://www.stuff.co.nz/national/92961550/highlevel-alarm-could-have-alerted-three-fishermen-on-board-illfated-fv-jubilee
3. TAIC, MO-2012-202.
4. TAIC, MO-2007-202.
5. TAIC, MO-2008-202.
6. TAIC, MO-10-204.
7. TAIC, MO-2010-206.
8. TAIC, MO-2017-201 and 202.
9. 'Timaru cruise ship crash puts safety at all ports under the spotlight', https://www.stuff.co.nz/business/112270886/timaru-cruise-ship-crash-puts-safety-at-all-ports-under-the-spotlight; TAIC, MO-2017-204.
10. 'Rena may become NZ's largest shipwreck', https://www.rnz.co.nz/news/rena-grounding/88116/rena-may-become-nz%27s-largest-shipwreck.
11. TAIC, MO-2011-204.
12. 'Final Report on Grounding of MV Rena', http://www.scoop.co.nz/stories/PO1412/S00250/final-report-on-grounding-of-mv-rena.htm
13. 'MNZ responding to incident off the coast of Tauranga, 5 October 2011, 10:45am', https://www.maritimenz.govt.nz/public/news/media-releases-2011/20111005b.asp
14. 'Rena grounding update - notices issued, 6 October 2011, 3:30pm', https://www.maritimenz.govt.nz/public/news/media-releases-2011/20111006f.asp
15. 'Stricken ship crew ashore after mayday call', http://www.stuff.co.nz/environment/5763630/Stricken-ship-crew-ashore-after-mayday-call
16. TAIC, MO-2011-204.
17. 'Worst ever environmental disaster', http://www.nzherald.co.nz/nz/news/article.cfm?c_id=1&objectid=10758195
18. 'Rena: Shipwreck that shook a nation', https://www.nzherald.co.nz/nz/news/article.cfm?c_id=1&objectid=10759216
19. 'New Zealand: Dangerous Goods Containers Removed from Bow of Rena', https://worldmaritimenews.com/archives/45369/new-zealand-dangerous-goods-containers-removed-from-bow-of-rena/
20. 'Rena oil spill: Workers brace for death of ailing ship', http://www.nzherald.co.nz/nz/news/article.cfm?c_id=1&objectid=10759860
21. 'Rena's stern gone from reef', http://www.odt.co.nz/news/national/204218/renas-sterngone-reef
22. R Schiel, PM Ross & CN Battershill, 'Environmental effects of the MV Rena shipwreck: cross-disciplinary investigations of oil and debris impacts on a coastal ecosystem', *New Zealand Journal of Marine and Freshwater Research,* 50:1 (2016), p. 2.
23. 'One small step for the 'Bay of Empty'...', https://www.newsroom.co.nz/2018/06/20/126202/bay-of-empty
24. 'Rena disaster hurts National', http://www.stuff.co.nz/national/politics/5786118/Rena-disaster-hurts-National-site
25. 'Rena Recovery' issue 21 (August 2015) http://www.renarecovery.org.nz/media/26245/rena-recovery-newsletter-issue-21-august-2015-formatted.pdf

Index

A
Abel Tasman 48, 49, 239
Achilles, HMS 138, 141
Adams, Captain 38
Adams, Captain William 89, 90
Addenda 238
Adsteam Port Services 196
Agnes 15
Agnew, Captain 107
Ahipara (Northland) 17
Akaroa 169, 177, 210
Alam Seri 228
Albatross, SMS 90
Albion 77
Alexander Newton 223, 237
Alexandra (lifeboat) 71
Alligator, HMS 18
Alpha 89
Anatoki 226–27
Anchor Shipping & Foundry Co. 141, 167
Anderson, Elizabeth 88
Anglo-Norman 238
Anjou 80, 84, 238
Anson, J. 178
Antiocco Accame 75
Aorere 103, 222
Aotea 238
Arahura 208, 209, 211
Aramoana 200, 204
Aranuia 230, 238
Aratere 211
Aratika 171
Archibald, Captain W.D. 50
Areituru, see Danger Reef
Ariadne 130, 132, 237
Argyle 109, 114
Aryan 238
Ashby, Ted 163
Asiatic Triumph 191
Assaye 237
Assistance 00
Astrolabe Reef 223, 225, 229, 232, 233, 239
Auckland 170
Aurora 236
Austin, Captain Arthur Henry 4

B
Babot, Captain Edwin 59
Badger, Charlotte 15
Baker, Keith 103, 104
Baker, W. 111
Barker, H.L. 178
Banshee 63
Barrett Reef (Wellington) 4, 171, 172, 189, 190, 202, 205, 239
Bay Fisher 185, 187
Bay of Islands 14, 15, 218, 236
Belich, James 18
Bellinger 90
Ben Avon 222, 238
Ben Venue 65, 66, 69, 71, 237
Bertha Dolbeer 238
Bettison, Andrew 172
Bennett, Captain H.J. 94
Billinghurst, K.D. 178
biosecurity 26, 28
Black, Captain Felix 153
Black Cat 210
Blacklaw, Lieutenant D.W. 141
Bluff 13, 26, 28, 107, 197, 215, 222, 225, 228, 239

Bluff Coastgaurd 226
Bodewes shipyards 184
Bollons, Captain John 20
Boojum 237
Boyce, G.J. 178
Boyd 15, 17, 18, 214, 218, 222, 236
Bradley, Temporary Lieutenant Peter 92, 93
Bradstock, Mike 218
Breeze 176, 236, 238
Brett, Sir Henry 99
Bridge, Gerald 122
British King 237
British Sceptre 100
Broughton, Betsey 17
Bruce, Captain David 187
Buffalo, HMS 236
Bull Rock 154, 238
Burnett, Commodore William Farquarson 33, C1
Burnett, Harry 179
Button, Peter and Clive C16

C
Calliope, HMS 237
Calm 175–76
Cambodia 18
Campbell, Alfred 88
Campbell, Julia 88
Campbell, Duncan 88
Campbell, Edward 88
Campbell, Lillian 88
Campbell, Muriel 88
Campbell, Rev. Thomas 88
Canterbury Steam Shipping Co. 175
Cape Horn 181, 182, 184
Cape Ortega 177
Capella Voyager 28
Capitaine Bougainville 105
Carey, Captain M. 38
Caroline 63
Cargill, Captain William 34
Casey, Gerald 184
Cashman, Tim 141
Castaway on the Aucklands (book) 79
Castaways of Disappointment Island, The (book) 79
Castlepoint (Wairarapa) 23, 29
Catlins River (Otago) 35, 36
Caughey, Fred 82
Caunce, Captain Arthur 150
Cecilia Sudden 116
Charlotte Jane 236
Chatfield, Captain H.W.H. 110
Chatham Islands 18, 26, 144, 166, 236, 238, C5
Chiou, Captain R.Y. 191, 192, 194
City of Auckland 213, 237
City of Cashmere 65, 66, 237
City of Dunedin 236
City of Perth 66, 69, 71, 237
Clarkson, Victor 184
Claymore 90, 92, 141
Clayton, Captain M.T. 114
Clutha 34
Clyde 237
Cobar C5
Collett, D. 184
Collis, William 65
Colonial Sugar Refining Company 7
Columbus New Zealand 239
Compadre 84, 237
Constance Craig 238
Coptic 76, 95
Cook, Captain James 14
Cospatrick 99–100, 235

Costa Concordia 232
County of Ayr 238
County of Carnarvon 237
Cox, Captain E.A. 47
Cox, Captain Frederick William 49
Cox, Greg 28
Crabtree, Captain Derek 103, 104
CTC Lines 206, 209
Culbert, Captain William 57, 58, 60
Cyrena 238

D
Dallam Tower 77
Daniels, J. 47, 51
Danger Reef (Otago) 14, 75, 238
Darroch, Captain R. 189
Davis, Thomas 17
Davies, Captain 77
Davies, Captain C.M. 175
Day, Bill 82
Defender 102, 103
Delaware 236
Department of Conservation (DOC) 28, 84
Derry Castle 78, 84, 237
Deuchrass, James 89
Devon 150–53, 154, 189, 238
Dickie, Captain 88
Doak, Wade 15, 214
Dobie, Gavin 163
Doctor Hook 239
Dog Island (Foveaux Strait) 26, 239
Dong Won 239
Doris 238
Doubtless Bay (Northland) 14, 218, 236
Downie, Captain 38
Dragon 18
Drake, Captain W.D. 176
Driver, HMS 239
Driver, Richard 19
Duchess 150
Duke of Sutherland 65, 66, 237
Duke of Edinburgh 66
Dunajtschik, Mark C14
Dundonald 82–84, 238
Dunedin, (tug) 25, 190
Dunedin, (ship) 75–76, 237
Dunedin 18, 19, 25, 32, 35, 44, 47, 48, 71, 76, 87, 89, 90, 113, 115, 128, 130, 131, 175, 178, 190, 214
Dusky Sound (Fiordland) 14, 218, 236, 238
Dutton, Peter C14
Duus, Captain C. 13

E
Eastern Honor 28
Easy Rider 239
Eastland Port Ltd 197
Echo 161–65
Eckford, Captain T.S. 165
Eckford, Charles 163
Edge, Captain Thomas 181
Edisto, USCGS 103
Edwin, Captain A.M. 150
Elderslie (schooner) 65
Electra 78
Elginshire 71, 237
Elingamite 217, 238
Elmslie, Captain Alexander 99
Elmslie, Henrietta 99
Emerald 238
Emile 61, 64, 237
Emu 56
Emu (ferry) 236

245

Emulous 55, 63, 64
Endeavour 14, 218
Endeavour, HMS 236
Endeavour, HMNZS 195
Enterprise 237
Edwin Fox 215
Escott–Inman, Reverend H. 79
Evelyn 237
Exxon Valdez 234

F
Fairchild, Captain John 20
Fairy Queen 56, 237
Farrell, George 122
Favourite 87, 89, 90, 236
Fiery Star 236
Fifeshire 236
Finch, Fanny 88
Fishwick, Captain A.J. 158
Flat Rock (Hauraki Gulf) 7
Fletcher, Charles 184
Florence 46, 236
Flying Foam 32, 32
Flying Scud 81, 213
Foremost 17 141
Forges, Henry 81
Forrest Hall 238
Fowler, Captain E.H. 23
Foxton 238
Foxton C2
Francie 225, 239
Francis, Ian 117, 214
Franklin Belle 246
Freke, Eric 132
Friendship 65
Frost, James 88

G
Gairloch 4, 90, 128
Gale, HMS 143, C10
Garrick, J.D. 179
Gee, Brendan 172
General Grant 7, 79–82, 84, 214
Geo H. Scales 209
Gisborne 28, 148, 149, 167, 194–96
Gisborne Sheepfarmers' Frozen Meat & Mercantile Company Ltd 166, 167
Golden Age 90
Golden Bay 208
Goldseeker 46
Goode, Sir John 41, 47
Gordon, Captain H. 183
Gordon, Patrick 179
Gothic 107
Governor Wynyard 236
Grafton 79–84, 213–15
Gray, Captain W.A. 48, 49
Great Barrier Island 109, 110, 114, 155, 232, C4, C7
Greenpeace 133–35
Greymouth 19, 20, 39, 41, 45, 47–51, 53, 182, 183, 226, 237, 238, 239
Grieve, Captain Alex 179
Grindrod, Lizzie 111, 113
Guard, Betty 18, 236
Guard, John 236

H
Halcione 237
Hampson, Eric 187
Hampson, George 202
Hancox, Captain John 176, 177, 178
Hand, George 35

Hanjin Bombay 227
Hannam, Ada 121, 122
Hansby, Captain John 128, 129
Harrier C1
Harriett 18, 236
Harding, W.H. 178
Harris, George 81
Hart, Captain J.M. 149
Hart, William 140
Hastie, F. 114
Hauroko 196, 197
Hautapu 132, 133, 242, C11
Hawea (1875) 47, 48, 127–30, 237
Hawea (1897) 47–48, 238
Hawea (trawler) 239
Hayward, Captain Bertram 155
Hector, HMS 138
Heemskerck 14, 236
Hempstalk, A.J. 103, 104
heritage value of shipwrecks 217
Hikitia 133, 173, 206, 222
Himalaya 78
Hinemoa 20, 21, 79, 83, 84, 200
Hobley, G.E.R. 141
Hokitika 41, 43, 44–47, 236
Holdsworth, Charles 47
Holm, Captain John 178
Holm Shipping Company 103, 144, 168
Holmbank, see *Turihaua*
Holmburn 103–105, 177, 178
Holmglen 175–80, 239, C15
Holmpark 206
Holmwood 144, 145, 239
Hope 63
Howells, F.F. 145
Howie, Captain Alexander 92, 93
Huddart Parker 32, 137, 189, 190
Hydrabad 213, 237, C2

I
Inconstant 216, 236
Ingham, R.C. 187
Ingram, C.W.N. 7, 127, 214
Inkster, Johnny 180
Invercauld 78, 79, 81, 82, 214, 236
Inverell, HMNZS 133, 183
Iona 237

J
Jamison, Captain Don 207, 210
Jane Anderson 61
Jane Douglas 43
Jean Bart 18
Jenkins, Michael 184
Jody F. Millennium 28, 194–96, 225, 239
John Wickliffe 236
Johnson, Captain Edmund 35
Jones, Geoffrey 184
Joseph Craig 238
Jubilee 225–26, 239

K
Kahawai, HMNZS 187
Kahu 39
Kaiapoi 39, 182
Kairaki 238
Kaitawa 133, 175, 180–84, 239, C12
Kaiwarra 239
Kakanui (North Otago) 41, 63
Kakanui (steamer) 84, 237
Kapanui 90, 92
Kapiti 141
Kaponga 40, 41, 48–49, 238

Karaka 102, 103, 150
Katie S 116
Katoa 156
Keera 195, 196
Keith, Captain 'Tup' Cosmo 179
Kekeno 83
Kells, Captain H.J. 138
Kemphaan 239
Kennedy, W.A. 51
Kent 211, 239
Kerr, Reginald 140
Kerry, Thomas 130
Kia Ora 239
Kilminster, Ronald 103
Kingston, Thomas 89
Kini 38
Komata 49, 145
Komet, KMS 14, 145, 239, C9
Konini 182, 238
Koputai 25
Korowai 177
Kotiti 238
Kotuku (steamer) 238
Kotuku (trawler) 239
Kuaka 167
Kulmerland, KMS 144, 239
Kupe 191, 209

L
L'Austral 228
Lady Elizabeth II 191, 209, C16
Laird, Captain Jock 143
Lange, David 134
Lastingham 237
Lauderdale 238
Leach, Captain 41
Leahy, Paddy 176
Leda Maersk 228
Legacy 225
legislation Enquiry into Wrecks Act (1863) 21
    Harbour Regulations Ordinance (1842) 236
    Historic Places Act (1993) 220
    Marine Board Act (1862) 236
    Passengers Act (1855) 34
    Shipping and Seamen's Act (1877);
    Shipping and Seaman Act (1903) 23
lighthouses Cape Egmont 175, 176
    Pencarrow Head 120, 150
    Taiaroa Head 19
    Waipapa Point 21, 23
lighthouse tenders 20, 21
Lillie Durham 65
Lindbergh, Captain Thoby 184
Little Mermaid 210
Lizzie Bell 237
Loch Lamond 238
Loch Long 238
Locke, Elsie 34
Locker-Lampson, Steve 117, 214
Lodestar Forest 226–27
Longworth, William C5
Lord Worsley 236
Lott, Captain Rod 170–72
Loughlin, Captain William 82
Lucena, Craig 170–72
Lena 163, 239
Luke, William 122
Luly, Rodney 202
Lumley, A.J. 113
Lynn, D. 122
Lyttelton 169
Lyttelton 26, 103, 118, 185, 222

M

MacDonald, Captain 39
MacDonald, Captain C. 69
MacGregor 90, 92
MacKenzie, Captain Josiah 101
Maclaren, Alexander 81
MacNeil, Captain James 94
MacQuaid, Annie 111, 115
McCormick, David 122
McDonald, Captain Col 47
McDonald, Charlotte 111, 115
McDonald, Henry 100
McGregor, Alexander 90
McHardy, F. 187
McIntosh, Captain John 110–11, 113–116
McKay, Allan 179
McKenzie, George 59
McKinlay, James 89
McKinnon, J. 116
McLachlan, A.A. 95
Mclean, Captain 60
Mclean, George 128
McLellan, David 82
McMeckan Blackwood & Co. 44
McPherson, J.H. 187
McTaggart, David 133, 135
Mafart, Alain 134
Makepiece, Captain Harry 154
Malthus, R. 177
Manaia 238
Manapouri 109, 113
Manchester 238
Manga, HMNZS 204
Manuka 213
Manukau Harbour 19, 25, 33, 104, C1
Manurewa 22
Maori   attacks on shipping 15, 18
    early contact with Europeans 15
    voyaging canoe traditions 13
Mapourika 47, 50–53
Maranui 175, 184–87, 239
Margaret Campbell 64
Margaret Galbraith 78
Maria 236
Maria Luisa 170–73, 239
Marine Department (later Ministry of Transport) 18–21, 23, 26, 31, 47, 49, 83, 96, 104, 162–64, 167, 176, 180
Maritime Archaeological Association of New Zealand 222
Maritime New Zealand (formerly Maritime Safety Authority) 26, 28, 211, 230
Marlborough 237
Marsden Point 28
Martin, Captain William 140
Masterman, Rowland 179
Matanaka 238
Matenga, Huria 236
Matthews, Ellis 121
Matoaka 78, 88, 89
Mattson, B.A. 141
May, Philip Ross 45
May Queen 239
Mayor Island C3
Mediterranean Shipping Company 28
Melbourne Steamship Company 39
Mesley, Pete 210
Mikhail Lermontov 7, 199, 206–211, 225, 229, 232, 234, 235, 239, C16
Milburn Carrier 208
Milburn Carrier II 225
Milford Lagoon (South Canterbury) 55
Mill, John 130
Miller, Captain James 144
Mills, Captain James 65
Mills, James 128–29
Milroy, Captain E. 183
Miranda C1
Mirrabooka 187
Moa (coaster) 102
Moeraki (North Otago) 60, 190–91
Moeraki Boulders 14
Molly Manx 228
Monk, G.B. 187
Montmorency 101, 236
Morley, Anne 17
Motiti Rohe Moana Trust 232, 233
Motutapere Island C13
Mumford, George 130–32
Murray, Captain A.H.G. 191
Murray, Charles 89
Musgrave, Captain Thomas 79–81

N

Napier 55, 101, 158, 163, 166, 229, 236, 237, 238, 239
Naylor, Captain Francis 118, 120–21, 123
Nelson 9, 50, 53, 118, 127, 137, 138, 173, 213, 214, 236, 237
Nelson 41
Neptune 162
New Plymouth 19, 28, 50, 55, 65, 128, 175, 176, 236, 237
New Zealand Company 18
New Zealand Merchant Service Guild 24
New Zealand Shipping Company 143, 144, 154, 237, C9
New Zealand Shipwrecks (book) 6, 7, 214
Newman, W. 128, 129
Ngahere 48, 238
Niagara 96, 137–41, 209, 234, 239, C10
Northern Monarch 238
Northern Steam Ship Company 4, 90, 184
Northcraft, H.W. 114, 115
Northland Harbour Board 105, 182
Northumberland 237
NZ Fisheries Ltd 133

O

Oakton, Richard 180
Oakura Reef (Taranaki) 4
Oamaru 7, 19, 24, 55–61, 63–65, 66, 75, 93, 130, 176–79, 236, 237
Ocean Chargers Ltd 191
Ocean Wave 64
Ohinemuri 22
Ohau 237
Olsen, Captain Athol 202
Opua 23, 238
Okta 13
Onehunga (see also Manukau Harbour) 25, 33, 50, 128, 169
Opouri 47, 48, 238
Opouri Shipping Company 47
Orion, KMS 138, 140, 141, 143–45, 239, C9, C10
Orpheus, HMS 31, 33, 234, 236, C1
Osprey, HMS 236
Otago (ketch) 56, 59–61
Otago (steamer) 237
Otago (tug, 1956) 25
Otago (tug, 2003) 25
Otago Harbour Board 25

P

Pacific Charger 191–94, 225, 226, 239
Pacific Chieftain 195
Paiaka 213
Panepane Point 229
Parahaki 182
Parramatta 14
Pateke 103–105
PeeJay V 239
Pemberton, A. 178
Penguin 23, 50, 109, 117–123, 222, 238
Penman, Frank 13
Perth 39, 48, 238
Peters, Jabez 83
Petersen, Captain Ned 161–62
Pereira, Fernando 134
Picton 53, 75, 118, 161, 207, 211, 215
Pleiades 237
Ponsford, Captain Steve 209
Poranui 185, 187
Port Bowen 149, 158–59, 239, C8
Port Chalmers 18, 36, 65, 75, 77, 78, 99, 110, 115, 176, 190, 215, 222, 228, 236, 237
Port Elliott 149, 157–58, 238
Port Gisborne Ltd 196
Port Kembla 137, 138, 238
Port Otago Ltd 25
Prebble, Hon. Richard 209, 210
Premier 64, 65
President Felix Faure 79, 84, 238
Pride of the Yarra 87–90, 236
Prieur, Dominique 134
PrimePort Timaru 228
Progress 238
Providence 14, 236
Puriri, HMS 141, 239, C10

Q

Queen Bee 237
Quickcat 210, 239

R

Rainbow, Captain E.F. 180
Rainbow Warrior 127, 133–35, 223, 239, C13
Randolph 236
Rangatira 200, 206, 238, 239
Rangi 25, C6
Rangitane 144, 145, 239, C9
Rata, HMNZS 92, C7
Raynal, Francois 79, 81
Rena, MV 148, 223, 225, 229–34, 239
Rona 7
Ronga, see *Wairau*
Regnaud, Captain Edward 'Joe' 177, 178
Reynolds, William Hunter 87
Richardson & Co. 39, 154, 158, 163, 166, 167
Rio Loge 123, 238
Ripple 23, 238
Robert and Betsy 57
Roberts, J.C. 187
Roberts, Mary 88
Robertson, Captain Hector Gordon 199
Rosella 45
Rose-Noelle 235, 239
Ross, Captain 34
Rotomahana 38, 113
Rough, Captain David 236
Rowlands, Captain Henry 65, 66

S

Saint Jean Baptiste 13, 14
St. Vincent 237
San Cuvier 239
San Domenico 218
Sankichi Maru No. 18 239
Sariska, MSC 26

Scotsman 61, 64, 237
Seabourn Encore 228
Seafresh 1 26, 28, 239
Seawitch 7
Seed, William 31
Selco Marine Salvage 191
Seok, Captain Joe Dong 196
Sherlock, Captain George 184
shipwrecks biosecurity concerns 26–29
    early Maori/European interaction 13–18
    dangers of early coastline 18–19
    heritage value 213–223
    legal processes 20–23
    Maori legends 13
    naming of sites 213
    safety regulations 23–25
Shipwreck Relief Society of New Zealand (later New Zealand Shipwreck Welfare Trust) 32
Skerrett, Charles 132
Sir Francis Drake 43, 45
Sewell, Captain William 56–57, 60, 76
Smith, Bruce 103
Smith, Robert James 172
Sommerville, Charles 88
Sorensen, Hanne 134
Soundsgood 171, 172, 243
South Port 197
South Sea, HMNZS 92, 239, C7
Southern Salvor 28
Southpac 29
Spence, Captain Robert 89
Spirit of Resolution 25, 227, 239
Spirit of the Dawn 84, 237
Star of Canada 148–49, 238
Star of Tasmania 56–59, 236
Star of the South 45
Stately 61
Stephanischlev, Stephan 207, 210
Stewart Island 81, 84, 165, 236, 237, 239
Strang, Robert 38
Strathallan 34
Strathfieldsaye 34
Strode, A.R.C. 34
sub-Antarctic islands 20, 21, 79, 80, 81, 84
Sumi Maru 192
Sumner Volunteer Lifeboat Brigade 237
Sundgren, Paul 172
Surat 35, 36, 213, 237
Surville, Captain Jean-Francois de 13, 14, 218
Suva 48, 237
Sword, Captain Cyril 202
Sydney Express 169–73, 239
Symphony of the Seas 228

T
Taharoa Express 28
Tai Ping 194–97, 225
Taiaroa 237
Taiaroa Head (Otago Harbour) 19, 177, 190, 237, C5
Taioma 94, 176
Tairea (ancient voyaging canoe) 14
Takapu, HMNZS 177
Tanea 169, 176, 177, 179
Tapuhi 94, 202
Taranaki 87, 93–96
Tarihiko 208, 209
Tararua 7, 21, 23, 214, 237
Tarapunga, HMNZS 177
Tarlton, Kelly 214, 217, 218
Tasman, Abel 48, 49, 239
Tasman Express Line (TEL) 170

Tasmania 56, 217, 237
Taupo (wreck) 220, 237
Taupo, HMNZS 128, 208, C2
Tauranga 237
Tauranga 185, 196, 207, 220, 223, 225, 226, 227, 229
Taylor, Stephan 170
Te Ara 15
Te Pahi 17
Teer, James 80, 82
The Portland 92, 204
Thom's Rock (Cook Strait) 120
Thomas, Captain Jeane 105
Thomas and Henry 60
Thompson, Captain John 15
Thompson, Captain William 60
Thompson, Peter H. 115
Three Kings Islands 238, C4
Tiger III 210, 239
Timaru 56, 65–71, 169, 177, 178, 180, 185, 196, 213, 228, 236, 237
Timaru Landing and Shipping Company 69
Titirangi 194
Todd, Captain John 34
Todd, Captain D.M. 94
Toia 92, 159, 164, 191, 209, C7
Tokomaru Bay 15, 167
Tongariro 149, 154–55, 158, 238
Toogood, Captain 34, 35
Tora (Wairarapa) 23
Torea 226
Torrey Canyon 234
Towson, John Thomas 76
Traill, Roxby & Co. 60
Transport Accident Investigation Commission (TAIC) 26, 172, 225, 227, 228, 229
Treneglos 71
tugs 25
Tui, HMNZS 183, 218, 223
Turihaua (tug) 166–68, 194
Turakina 78, 143, 144, 237, 239
Tutty, Eric 179, 243
Twentyman, Chief Inspector George 204
Tyrone 149, 157, 238, C5

U
Ulimaroa 32
Union East 239
United Brothers 65
United Salvage Proprietary Ltd 141, 206
Upton, Captain H.L. 145

V
Ventnor 238
Venture 177
Verne, Jules 79
Victory 35, 236
Victory Beach (Otago) 34
Viggo Hansteen 189, 190
Vire, FNS 36, 237
Vistula 63
Vrobyov, Captain Vladislav 207

W
Wahine (1913) 92–93, 239
Wahine (1966) 7, 13, 150, 189, 192, 194, 199–206, 210, 214, 222, 239, C7, C14
Wahine Point C5
Waihora 113
Wahlstedt, Captain Thorsten 187
Waikare 238
Waikouaiti 26, 239

Waipapa Point (Marlborough) 9
Waipapa Point (Southland) 21, 23, 237
Waipiata 87, 93–96
Wairarapa 109–116, 213, 237, C4
Wairau 161–63
Wairau bar (Marlborough) 163, 165
Wairuna 137, 238
Waitaki 65
Waitaki River (North Otago) 130, 131, 237, 239
Waiweranui Pont (Taranaki) 175, 176
Wakatu 9
Wakatu Shipping Company 9
Waller, William 128
Walton, J. 187
Wakakura, HMNZS 208
Walara-K 226, 239
Wanderer 66, 127, 237
Wanganella 4, 141, 189, 190, 239
Wanganui 36
Whanganui 41, 53, 102, 158, 159, 162, 176, 222
Wareatea 38, 75
Water Nymph 56, 57, 59, 236
Watson, R.G. 187
Webster, Mel 172
Webster, Pat 172
Wellington 9, 18, 19, 20, 23, 26, 48–51, 79, 84, 92–94, 95–96, 102, 107, 117–18, 120, 123, 132–33, 143, 150, 153–54, 158–59, 162–64, 166–67, 169, 176, 189, 191, 199, 204, 207, 209, 211, 215–16, 218, 222–23, 236
Wellington Harbour Board 205
Wellington (ship) 75
Wellington (steamer) 90
Wellington, HMNZS 218
Wellington Express 170
Westland (tug) 39, 41, 48
Westport 25, 48–51, 53, 180, 182, 237, 238
Weyher, Captain Kurt 138, 140, 144
Whangape bar 22
Whangarei 25, 141, 185
Whangaroa Harbour (Northland) 218
White, Charles 114
White Swan 236
Whorlow, D. 178
William Bryan 236
William Denny 236
William Miskin 88, 89
Williams, Captain Henry 176
Williams, Royce 184
Willis, Captain Stewart 130
Wilson, Captain Frederick 89
Wiltshire 149, 155–56, 238, C7
Wimmera 137, 138, 238
Winders, J.C. 179
Wolf, SMS 137, 138, 140, 238
Wolgast, A.J. 178
Wonga Wonga 33
Worth, Russell C14
Wright, John 183, 184

Y
Young, Duncan 59

Z
Zacharia, Joseph 123
Zeehaen 14, 236
Zuleika 237